From Basin to Peak:
An Explorer's
Companion to the
Colorado-New Mexico San Juan Basin

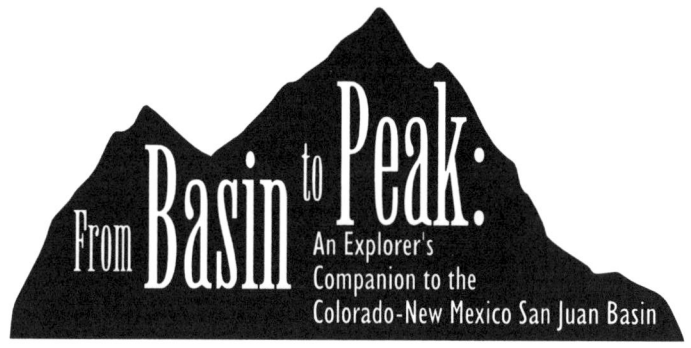

From Basin to Peak:
An Explorer's
Companion to the
Colorado-New Mexico San Juan Basin

Wesley M. Howe

Texas Tech University Press

This book was set in ITC Garamond and printed on acid-free paper that meets the guidelines for permanence and durability of the committee on Production Guide- lines for Book Longevity of the Council on Library Resources. ∞

Design by Melissa Bartz

Printed in the United States of America

Library of Congress Cataloging-in-Publication Data
Howe, Wesley M., 1933-
    From basin to peak : an explorer's companion to the Colorado-New Mexico San Juan Basin / Wesley M. Howe.
      p.   cm.
    Includes bibliographical references (p.   ) and index.
    ISBN 0-89672-395-X (alk. paper)
    1. San Juan Basin (N.M. and Colo.)—Guidebooks. I. Title.
F782.S19H69   1998
917.89' '82—dc21   98-14318
                                                           CIP

98 99 00 01 02 03 04 05 06 / 9 8 7 6 5 4 3 2 1

Texas Tech University Press
Box 41037
Lubbock, Texas 79409-1037 USA

800-832-4042

ttup@ttu.edu

Contents

We knew we had found the garden spot of the world for farming and stock raising. There were no laws to break and the life was a rosy dream.

George W. Coe, rancher and gunfighter,
recalling his 1878 impression of the San Juan country

Preface

To learn about the San Juan Basin and how it displays the American West, you could amass a small library. This book is a summary of what you would find in such a library. Along with an abstract of the area's natural and man-made feature and cultural history, it includes obscure facts and engaging anecdotes to help you better appreciate the basin and the West.

As I gathered information and stories for this book, I soon realized the basin is more than just a part of the West, it is an *example* of the West, and an exciting one. It is a metaphor, the West in miniature.

Look for a history or reference book elsewhere about the San Juan Basin and you won't find one. Yes, you can find lots of Colorado and New Mexico histories, but they stop at the state lines. The basin's history, like its culture, doesn't stop at a state boundary; it flows across the Colorado–New Mexico border. And in books about the broader West, the San Juan Basin gets lost.

You can find lots of publications about *pieces* of the basin—Durango, Silverton, Montezuma County, Mesa Verde, railroads, mining, ancestral Puebloans—if not in the bank of tourist brochures, surely at libraries and book stores. This work puts those pieces together.

This is neither a typical tourist advisory nor a work of propaganda to convince you, if you don't already live in the basin, to go there and spend money. Nor, at the other extreme, is it a definitive history. It is a collection of information and anecdotes of general interest whether you be a tourist, prospective resident, history buff, or general reader. You will find such a collection in no other publication.

No single work should try to be a comprehensive reference on everything in the basin and the American West, nor should it attempt to cover completely each of its selected subjects. In choosing topics for this book and deciding what they should embrace, I have asked: Are they useful? Are they interesting? Do they present an example? Will they help you better know the San Juan Basin and the American West?

By arranging the articles alphabetically, I have made it easy for you to select topics at random. To pursue a subject, turn to related entries mentioned in the text or listed in Appendix A. To explore a subject in depth, note the bracketed numbers at the end of an article and refer to the sources. Many are readily available at libraries and book stores.

As a visitor, you can turn to Appendix B and use the suggested itineraries to explore the basin. But the entries in this book are not a

Preface

substitute for the particulars available from managing agencies, visitor centers, chambers of commerce, and tourist-oriented businesses.

For a chronology of the basin, review the introductory timeline, which you may want to consult as you read the entries. For while this work is more than a history, many of its subjects relate to the basin's past.

A good road map will help you enjoy much of the text. Any good map you like will do. Since most maps tend to stop at state lines, you may need more than one to overcome the artificial division that runs through the basin.

So much for the gist of this book. Now, should you want to know if the region has grizzly bears, get briefed about the area's wildlife areas, learn about Western author Louis L'Amour's affinity for the basin, or find out how Deputy Sheriff James J. Sullivan ended the career of gunfighter Ike Stockton, read on.

In writing this encyclopedia I have relied heavily on secondary sources. I am, therefore, indebted to all the writers, compilers, and historians who did the spade work that made this work possible. Their names appear as authors in the list of sources.

I am grateful to Duane A. Smith, director of the Center of Southwest Studies, Fort Lewis College, for his encouragement and to Catherine Conrad, his assistant. The *Farmington Daily Times* readily granted access to its archives, and its librarian, Lynette Chilcoat, provided assistance. Several museum curators helped locate photographs: Robert McDaniel and Charles DiFerdinando at the Animas Museum, Durango; Beah Folk, Aztec Museum Pioneer Village; and Catherine Davis, Farmington Museum. Ann Bond with the San Juan–Rio Grande National Forest provided a selection of scenic and wildlife pictures.

I especially thank Marsha Fenner, who took time between her service at the Durango Public Library and other work to critique the text and offer suggestions. I acknowledge also the patience displayed by the members of the publisher's editorial staff in working with an amateur historian. And I am deeply indebted to my wife, Barbara, whose financial support and endurance helped make this work possible.

Introduction

Straddling the Colorado–New Mexico border just east of Arizona and Utah, the San Juan Basin depicts the essence of the American West. Each of its waves of immigrants, from the pre-Columbian Puebloans to the homesteaders, rubbed against those who had come before—and sometimes against other members of their own group. The basin's history, like the saga of the West, is largely the story of those abrasions.

Centuries ago natives migrated from the north and grew a civilization, only to be supplanted by yet other natives. Seeking territory and gold, Spaniards explored the basin. To the mountain streams the trappers journeyed to replace the exhausted beaver of the East. Then came the miners to the San Juan Mountains, some driven by Pike's Peak misadventures. Up the valleys to the foothill meadows rode the cattlemen from Texas, herding their longhorns. From the Rio Grande valley Mexican Americans brought sheep over the Continental Divide.

Then homesteading families rolled in from across the Great Plains. To the still-open range and the primitive saloons of the mining camps came rustlers and gunfighters, drifting north from the Lincoln County War. Soon an influx of merchants, preachers, teachers, and lawyers leavened the mixture.

Like its larger brothers that spanned the continent, the narrow-gauge railroad arrived to exploit the region, bring tourists, and mushroom the economy—only to become redundant when roads paved the way for motor transport. By the middle of the twentieth century the stockmen, farmers, and tourists were rubbing shoulders with seekers of energy—geologists, oil drillers, gas-line builders, and uranium prospectors.

The basin's terrain and climates, like its history, match the broader spectrum of that territory between the 100th Meridian and the Cascade Range known as the arid West, for its maze of high mountains slope to an expanse of desert laced with badlands.

But how shall we set the boundaries for this metaphor of the American West to which many refer, but few define, called the "San Juan Basin?"

Even geological descriptions can be ambiguous, for nature draws no boundaries. Much of the region lay under an ancient sea that left sediment to form its badlands. Other sections underwent geological turmoil as the earth uplifted, glaciers churned the crust, and water eroded its surface. The basin thereby became one of several among the ranges of the Rocky Mountains. By this definition it is a giant oval in northwestern New Mexico extending into Colorado until it reaches the mountains, spilling into Arizona, circling around Grants, New Mexico, then stopping short of Albuquerque. But geologists also refer to a central geologic basin, a much smaller bowl within the broader depression. A cultural description is more in line with this smaller basin.

Most regional businesses that say they are "Serving the San Juan Basin" have, or want, customers in Farmington, Durango, Cortez, Pagosa Springs, Shiprock, and every place between. But few would expect a call from Grants (New Mexico) even though it lies within the basin's geological description.

If confronted with no practical restraints, the economist might draw a basin that is different from both the geologist's and the businessman's. But economists rely on data collected for political entities; their limits are predetermined. Boundaries set for any purpose are necessarily arbitrary.

For this work the limits are a compromise of topography with political reality: it is that region in southwestern Colorado and northwestern New Mexico bordered on the north by the northern limits of Dolores and San Juan Counties, Colorado; on the west by Arizona and Utah; and on the east by the Continental Divide. The southern limit uses a line stretching from Arizona east along the southern border of New Mexico's San Juan County and extending east across Sandoval County to the divide.

The basin's 45,000 square miles are home to more than 150,000 people. Many residents are descendants of past generations of Native Americans, Hispanics, and Anglo settlers who have passed through it. But most are later immigrants drawn by the myth of the West, financial opportunity, or a lifestyle fantasy.

They live in a cultural whole (insofar as there is such a thing) split by the imaginary line separating Colorado and New Mexico. Whether old-timers or newcomers, the basin's residents view the division with a sense of reality and amusement. A February 13, 1926, editorial in the *Farmington Republican* suggested (presumably tongue in cheek) doing away with the boundary by splitting Colorado and New Mexico north-south instead of east-west. Then Denver and Santa Fe could vie for the capital of Colorado on the east, with Farmington and Durango drawing straws to be the center for New Mexico.

A 1996 quip among Coloradans about New Mexico drivers reveals friendly rivalry between those on opposite sides of the line: What do they do with New Mexicans who flunk their driving test? Make them drive with yellow license plates. (Yellow was the standard color for New Mexico's license plates.) No doubt the residents south of the border direct similar levity at their northerly neighbors.

Political manipulation divided the San Juan Basin. Social and economic verities, leavened with a sense of humor, keep it whole.

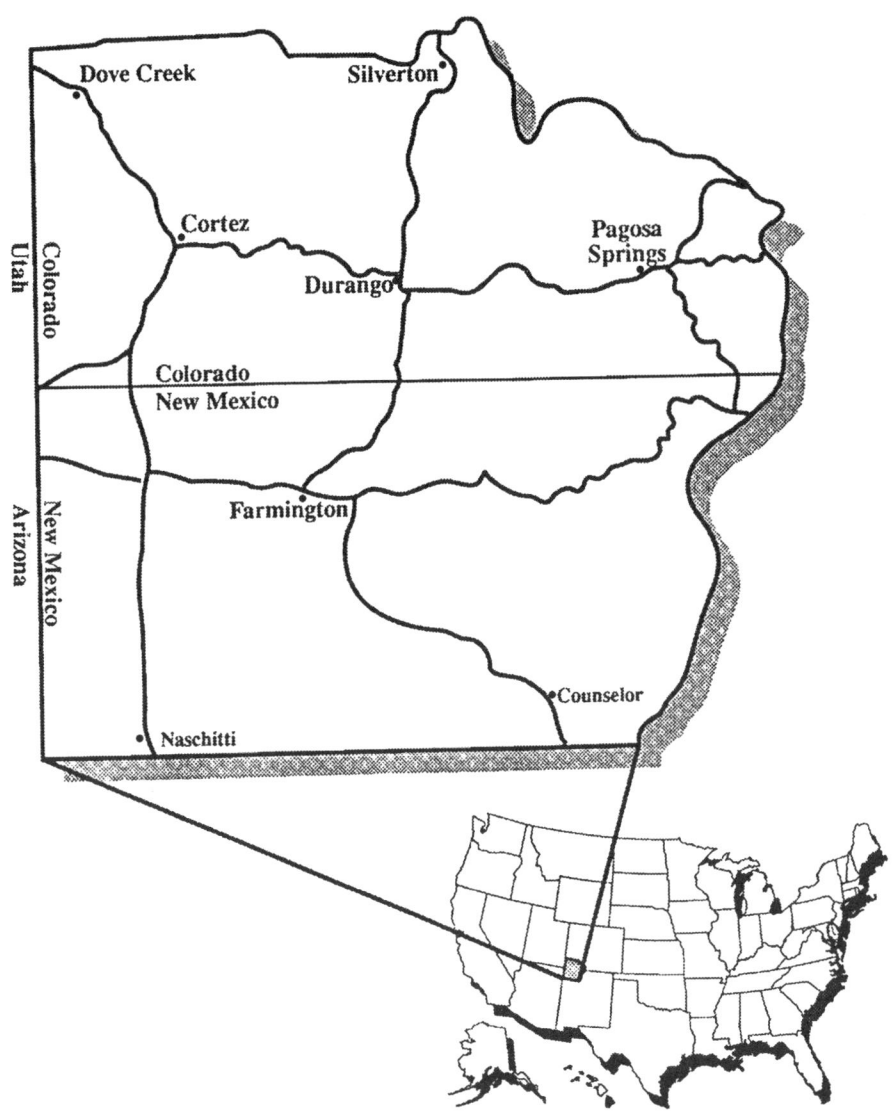

Dove Creek
Silverton
Cortez
Pagosa
Springs
Durango
Colorado
New Mexico
Farmington
Counselor
Naschitti

Utah
Colorado

Arizona
New Mexico

San Juan Basin

A Timeline of the San Juan Basin

Here's a capsule history of the West viewed through the window of the San Juan Basin. Some events are significant, others merely indicative.

pre-1400	Ancestral Puebloans, descendants of people who crossed an ice bridge from Asia 20,000 years ago, occupy the basin.
1400–1500	Navajos and Apaches migrate south from Canada along the east slope of Rocky Mountains and spread into the basin.
1540–42	Spanish explorer Francisco Vasquez de Coronado marches through New Mexico, south of the basin, in search of the Seven Golden Cities of Cibola; lays groundwork for Spanish claim of boundless territory including the basin.
1598	Juan de Onate begins Spanish colonization of New Mexico in lower Rio Grande valley. Missionaries begin campaign to uproot native religions.
1608	Spain sets up New Mexico (encompassing the basin) as a royal province.
1680	New Mexico Puebloans in Rio Grande valley revolt against imposition of European culture and religion, massacre and defeat Spanish colonizers.
1691–93	Diego de Vargas leads Spanish in reconquest of New Mexico.
1765–75	Juan Marie de Rivera and other prospecting Spaniards explore the basin.
1776–77	Francisco Atanasio Dominguez and Silvestre Valez de Escalante reconnoiter the basin looking for route from Santa Fe to Monterey.
1821	New (American) Spain declares independence from Old (European) Spain.
1822	William Becknell stimulates Anglo-American migration to Southwest by replacing pack train with wagons on the Santa Fe Trail running from Missouri to Santa Fe.

1824	New Spain forms Republic of Mexico, whose territory includes the basin.
1824–25	Mountain man Ewing Young scours San Juan River for beaver.
1830	Antonio Armijo and other Spanish traders establish Old Spanish Trail, running from Santa Fe through the basin to California.
1836	Texas secedes from Republic of Mexico and forms Republic of Texas.
1845	United States annexes Texas, sets stage for war with Mexico.
1846	U.S. Army colonel Stephen Watts Kearny marches to Santa Fe, proclaims New Mexico (including the basin) to be a possession of the United States.
1846–48	United States fights Mexico to secure annexation of Texas and fulfill America's "Manifest Destiny."
1847	Mormons establish a colony in future Utah Territory.
1848	By the Treaty of Guadalupe Hidalgo, Mexico cedes to United States the San Juan Basin as part of vast territory extending to the Pacific Ocean.
1849	Explorer John Charles Fremont, seeking a railroad route, meets disaster attempting winter crossing of the San Juan Mountains. Mormons propose State of Deseret, encompassing an extensive part of the West that includes the basin.
1850	United States and Texas adjust Texas boundary to its present dimensions. Congress divides Mexican cession into Utah and New Mexico Territories; sets boundary along 37th Parallel through the middle of the basin.
1860	Charles Baker leads prospecting party over Cinnamon Pass into San Juan Mountains seeking precious metals.
1861	Congress sets up Colorado Territory, using territorial border between Utah and New Mexico as its southern boundary and forever splitting the basin.

1862	Federals from Colorado, the "Pike's Peakers," drive Confederate Texans out of New Mexico; basin remains under Union control.
1863	Congress establishes Arizona Territory, removing it from Rebel-sympathetic (though Union-controlled) New Mexico; sets eastern boundary along same meridian as Utah-Colorado line, thus establishing the Four Corners and the basin's western boundary.
1863–64	General James H. Carleton and Colonel Kit Carson carry out scorched-earth policy and force Navahos to take "long walk" to Bosque Redondo concentration camp near Fort Sumner in southern New Mexico.
1868	Congress abandons Bosque Redondo experiment, establishes Navajo Reservation.
	Jicarilla Apaches end resistance.
1870s	Ferdinand V. Hayden and others carry out U.S. Geological Service mapping of the West.
1873	Utes cede mineral-rich area of San Juan Mountains; miners begin rushing in.
1874	Joseph Glidden invents barbed wire.
1876	Congress admits Colorado as thirty-eighth state.
1878	U.S. Army establishes Fort Lewis at Pagosa Springs to "protect and control" Southern Utes.
1879	Utes kill agent Nathan Meeker and others, kidnap women at White River Agency; the incident raises cries for removal of Utes to Utah. General Philip H. Sheridan visits basin on military inspection tour.
1880	Land in New Mexico between San Juan River and Colorado becomes public domain after six years of closure as an unused Jicarilla Apache reservation. Fort Lewis military post moves from Pagosa Springs to location on La Plata River.
1881	Denver and Rio Grande Railway arrives at Durango.
	Farmington vigilantes raid Durango seeking Ike Stockton and other outlaws wanted in New Mexico.
1882	Denver and Rio Grande Railway reaches Silverton.

1886	Army forces capture Geronimo, marking end of Native American hostilities in Southwest.
1887	President Grover Cleveland establishes present-day Jicarilla Apache Reservation.
1890	Rio Grande Southern Railroad builds to Rico and over Lizard Head Pass, providing a basin connection to Denver and Rio Grande Railway to the north. Congress enacts Sherman Silver Purchase Act, silver prices rise, mines and mills flourish.
1891	Army deactivates Fort Lewis.
1893	Congress repeals Sherman Silver Purchase Act, silver prices collapse, silver mines close, Rico and other silver camps become ghost towns.
1895	Congress establishes present-day Ute reservations.
1905	Denver and Rio Grande Railway builds to Farmington.
1906	Congress establishes Mesa Verde National Park.
1911	Zane Grey pens *Riders of the Purple Sage* near Dove Creek.
1912	Congress admits New Mexico as forty-seventh state.
1915	Jack Dempsey knocks down Andy Malloy in ten-round fight at Gem Theater, Durango.
1921	Oil syndicate drills basin's first commercial natural gas well near Aztec.
1922	Geologists find Hogback Oil Pool west of Farmington.
	Colorado River Basin states endorse Colorado River Compact, dividing water between upper and lower river basin states.
1935	Will Rogers stops in Durango a month before perishing in Alaskan plane crash.
1950	Farmington residents report UFO sightings.
1951	Oil boom starts population explosion.
1953	Uranium processing mill starts up in Durango.
1958–61	Bureau of Reclamation builds Navajo Reservoir.

1969	20th Century Fox films *Butch Cassidy and the Sundance Kid* with basin locations.
1978	Arizona prison escapee Gary Tison and gang rampage through Southwest, murder couple near Chimney Rock.
1981	Charles E. Bradshaw Jr. buys Denver and Rio Grande's Durango-Silverton line and names it the Durango and Silverton Narrow Gauge Railroad.
1990s	Basin experiences accelerated immigration of Californians, easterners, Texans, and others seeking lifestyle changes.
1994	Silverton's last gold mine closes, marking end to basin's mineral industry.

An Explorer's Companion to the San Juan Basin

Agricultural Science Center at Farmington. A branch of New Mexico State University, Las Cruces, this station develops plant materials and methods to improve production of such farm crops as alfalfa, corn, dry beans, small grains, onions, and other produce. Located southwest of Farmington, the center is also known as the New Mexico State University Agriculture Experimental Station.

Alcatraz, San Juan County, New Mexico. A Hispanic community settled in the late nineteenth century (and since disappeared) at Largo Canyon on the south side of the San Juan River near present-day Blanco. [3]

Allison, La Plata County, Colorado. Adjacent to Navajo Reservoir and resting near the abandoned route of the Denver and Rio Grande Railway, this community was settled in the early 1900s after the Ute Strip was opened to white occupation. It was originally called Vallejo, Spanish for "wide valley."

An early Allison settler related the story of a Ute entering her tent and taking a fresh cake. As he left, he gestured toward a pony and left it in payment for the baked goods.

Allison (elev. 6,213 feet) is eleven miles southeast of Ignacio on Colorado 151. [180]

Alpine Loop. A Scenic and Historic Byway that goes east out of Silverton through Animas Forks and returns to U.S. 550 north of Red Mountain Pass. The route is marked by the columbine emblem of the Colorado Scenic and Historic Byways Commission. [209]

Alpine Triangle Recreation Area. The Bureau of Land Management has so designated its lands east and northeast of Silverton. The area extends over the Continental Divide to Lake City. [149]

Amargo, Rio Arriba County, New Mexico. If you wanted to get off the Denver and Rio Grande Railway and take a stage up to Pagosa Springs for a hot bath in the 1880s, this was your station. But when Ed Vorhang filed a homestead on the property where the town was situated and decided to start charging rent, the merchants and others just moved a couple miles west to a place they called Lumberton. The post office followed in 1894.

Amargo faded as a town, although a few stockyards and loading chutes remained in the 1930s and a creek there still bears its name.

Amargo was about five miles east of Dulce on a road that is now U.S. 64. [91]

Anasazi. A term applied to some of the natives who lived on the Colorado Plateau before the fourteenth century. In connoting a specific tribe or culture, the term is not accurate, for the basin and much of the Southwest was home to several pre-Columbian cultures. To some Navajos the word means "ancient people" while for others it means "enemy ancestors." Some Native Americans and anthropologists suggest a more accurate and meaningful term: "ancestral Puebloans."

Anasazi Heritage Center. This depository interprets the Native American and other cultures of the Four Corners. Managed by the Bureau of Land Management, it has two million records and artifacts, descriptive exhibits, and films.

The museum is on Colorado 184 three miles west of Dolores. [151]

Anasazi Recreation Area. The Bureau of Land Management has given this designation to an area larger than 300 square miles in the northwest section of the basin. Lying west of Cortez, it encompasses McElmo Canyon, Hovenweep National Monument, and Lowry Pueblo Ruins. [149]

Ancestral Puebloans. Natives, commonly called "Anasazi," who lived on the Colorado Plateau from the Basket Maker Era (A.D. 1) into the fourteenth century. Although they are generally assumed to be the ancestors of some modern-day Native American tribes, why they migrated south after 1300 is a subject for both study and speculation. Causes that possibly contributed to their exodus include prolonged drought, warfare to control irrigated lands, and the attraction of alternate religions. They were artistic in their crafts, as revealed by basket and pottery

Cliff Palace at Mesa Verde National Park is among the most dramatic of the thousands of ancestral Puebloan remnants scattered throughout the San Juan Basin. Courtesy La Plata County Historical Society.

artifacts. And they were good at wresting a living from an inhospitable land. They used a variety of tools fashioned from stone, bone, and wood for hunting and farming. From early pithouse building, the Puebloans' house-building skills evolved to above-ground structures and cliff dwellings.

You can find remains of their buildings throughout southwestern Colorado and Utah and most of Arizona and New Mexico. As the home of Mesa Verde National Park and Chaco Culture National Historic Park, Hovenweep and Aztec national monuments, Pueblitos of Dinetah, Chimney Rock Archaeological Area, and other sites scattered about, the basin offers many examples of ancestral Puebloan architecture. [24, 118, 163]

Andrew's Lake State Wildlife Area. Located in the East Lime Creek drainage of the Animas River, this eight-acre lake lies within the San Juan National Forest. The Colorado Division of Wildlife and the U.S. Forest Service built it in the 1960s and continue to manage it. The division stocks the lake with trout. The lake's surface is at 10,744 feet elevation in the Montane vegetation zone.

Andrew's Lake is a half mile east of U.S. 550, eight miles south of Silverton. [250]

Angel Peak National Recreation Area. This place features a forty-million-year-old geological formation surrounded by badlands that are the dwelling place of sacred Navajo spirits. The Spaniards called it *nacimiento,* meaning "birthplace," and revered the feature's outspread wings.

The peak formed when ancient seas laid strata of mud capped with a thick layer of sand that turned to stone. After the seas left, the elements shaped the sandstone by washing or blowing away much of the underlying shale, leaving wrinkled badlands tinted by minerals with reds, yellows, lavenders, tans, grays, and browns. The top layer, called the San Jose Formation, survived erosion to provide the crown for Angel Peak, elevation 6,989 feet.

Operated by the Bureau of Land Management, the site is fifteen miles south of Bloomfield on New Mexico 44. San Juan 7175 leads six miles to the peak. [142, 152]

Animas City, La Plata County, Colorado. Two communities have had this name; none has it today.

The first Animas City was located near Baker's Bridge twelve miles up the Animas Valley from the center of modern-day Durango. Its settlers

Animas City, ca. 1880. As a farming and mercantile center, this community linked the San Juan mining districts with New Mexico suppliers. The community thrived until the Denver and Rio Grande Railway bypassed the town. Courtesy Fort Lewis College, Center of Southwest Studies.

were part of the San Juans' first mining rush, and it survived for only a year or so—from 1860 to 1861.

A second flock of gold seekers rushed into the San Juans in 1873 after government negotiators (and the army) persuaded the Utes to give up the area for mining. The fertile valley had no mines (there was only a nearby coal deposit), but merchants, traders, and freighters came to meet the miners' needs. Optimistic pioneers also settled on the west bank of the Animas River some ten miles south of the original Animas City. Incorporators filed a town plat in 1876. The community grew slowly to attain a population of 286 in 1880. Ranchers, farmers, and miners fed its economy. The army contingent that came in 1879 to protect the town from a perceived Ute threat, prompted by the so-called Meeker Massacre, helped assure prosperity.

The demise of the second Animas City began in December 1880 when officials of the Denver and Rio Grande Railway came to town. It was their practice to get concessions—land, rights-of-way, financial investment—from a community before honoring it with a depot. The town's trustees declined to meet the railroad's requests. With no favors forthcoming from Animas City, the Rio Grande's planners simply moved two miles south and Durango was born.

Animas City faded as merchants, professionals, and the newspaper scurried to set up shop in the new community. After nurturing the valley's agriculture and looking after the hardrock mining of the San Juans to the north, Animas City saw its growing prosperity taken away by upstart Durango. In the mid-1880s Animas City's optimism revived as a copper smelter started up, but Durango's smelter responded by doing a better job. Although the community revived after the initial shock of the railroad's snub, the town had a hard time financing its services. So the town leaders began exploring consolidation with Durango. The now several times larger Durango, seeing little advantage, turned down the idea. Finally, in 1947, serious negotiations led to a vote by the citizens of both communities in favor of merger.

Animas City (elev. 6,535 feet) became part of Durango. [84, 113]

Animas Forks, San Juan County, Colorado. This mining camp was at the end of the Silverton Northern Railroad where creeks merge to form the Animas River. It acquired a post office in 1875 and had its own, albeit unsuccessful, mineral processing mill. It lived at the mercy of the mining industry and faded away as the mines closed.

Animas Forks (elev. 11,160 feet) is thirteen miles northeast up Colorado 110 and San Juan 2 from Silverton. [97, 115]

Animas River. This stream begins where its two headwater creeks join at the ghost town of Animas Forks. It then passes Silverton, heads south where it has cut one of the longest and deepest gorges in the Rocky Mountains, meanders through the gentler segment of its valley, and continues past Durango to Farmington, where it joins the San Juan 110 miles from its source.

The river's sections below Silverton and downstream from Durango offer excitement for river runners. If built, the Animas–La Plata Project will reduce the river's flow below Durango. [104]

Animas–La Plata Project. An idea conceived by community boosters and dryland farmers in 1904 whose time, over ninety years later, has not yet come.

Consisting of a series of dams, reservoirs, pumping stations, piping systems, and ditches, this water project is designed to take water from the Animas River, where it goes under U.S. 160/550 at Durango, and pump it west uphill to a reservoir and the La Plata River watershed. From this higher elevation the water could flow to municipal, industrial, and agricultural users and to the Ute Mountain Ute and Southern Ute Reservations.

Arguments for this Bureau of Reclamation project have changed with time, reflecting the basin's growth and the desire to satisfy the tribes' water rights with the least inconvenience to other water users. [201]

Apache. See **Jicarilla Apache.**

Arboles, Archuleta County, Colorado. This community got started at the junction of the San Juan and Piedra Rivers within the Southern Ute Reservation in the early 1880s before the Ute Strip was opened for settlement. The community now rests next to the Navajo Reservoir that submerged the rivers' confluence in 1963.

Arboles (elev. 6,013 feet) is thirty miles southwest of Pagosa Springs via U.S. 160 and Colorado 151. [91]

Archuleta County, Colorado. Ancestral Puebloan ruins around Chimney Rock and Ute references to the county's hot springs are evidence of a history of Native Americans in this area.

Lumber companies hastened the region's European settlement when they built the Rio Grande and Pagosa Springs Railroad north from Lumberton in 1895. Five years later they also ran the Rio Grande, Pagosa and Northern Railroad to connect Pagosa Springs with the Denver and Rio Grande Railway at Pagosa Junction (sometimes called Gato). Several communities got started along the railroads to house lumber workers:

Talian, Alturas, Kearns, Lone Tree, Dyke, Nutria, Hatcher, Sunetha. Most disappeared after the forests were stripped. Others are now just names on the map.

As late as the 1960s, the lumber industry played a big role in the area, but since the opening of Wolf Creek Pass in 1916, tourism and recreation have also provided economic sustenance.

Archuleta County was split from Conejos County and established in 1885. County seat: Pagosa Springs; population (1990): 5,345; area: 1,364 square miles. [91, 208]

Armijo, Antonio. Spaniard who opened trade between New Mexico and California by helping blaze the Old Spanish Trail. His sixty-man party journeyed from Abiquiu, New Mexico, across the basin's La Plata and San Lazaro (Mancos) Rivers, circled north to avoid the Grand Canyon, and reached California's San Gabriel Mission in January 1830. [57, 112]

Armstrong, Neil Alden (1930–). An astronaut who spent a day at Electra Lake studying rock formations so he could compare them with what he might find on the moon. After becoming the first civilian to enter the astronaut training program and the first person to step on the moon, Armstrong became a professor of aerospace engineering at the University of Cincinnati in 1971 near his Wapakoneta, Ohio, birthplace. [95, 217]

Arriola, Montezuma County, Colorado. A choice agricultural locality settled shortly after the Montezuma Valley Irrigation Company promised water for the area in the 1880s.

Arriola is eight miles north of Cortez on U.S. 666. [50]

aspen. See **quaking aspen.**

Athapascan. An indigenous language of some Native Americans found primarily in Canada. The language's prevalence among the Apaches and Navajos supports the theory that their ancestors migrated south from Canada along the east slope of the Rocky Mountains. [118]

Aztec, San Juan County, New Mexico. Settled in 1880 as a farming community, Aztec was set by the territorial legislature to be San Juan County's seat when the new county was split from Rio Arriba County in 1887.

Two miles above where the Animas River merges with the San Juan, that designation upset the folks at Junction City who coveted the honor and the commerce it would bring. (The area is now part of Farmington.) Indeed, the process by which Aztec persuaded territorial governor E. C.

Ross to propose Aztec smelled of political skulduggery. His emissaries, sent up from Santa Fe to get the county started, had been suspiciously detained. And they never considered any place but Aztec for the county seat. Other communities, Junction City being the most vocal, demanded an election to determine which community would be the county seat. By a nine-ballot edge, that referendum gave the honor to Junction City.

Then a legal squabble began. Aztec refused to give up the county records. Only after territorial judge E. P. Seeds ordered the office moved to Junction City were the records surrendered. But in August 1892 the territorial Supreme Court reversed Seeds's order, giving the office back to Aztec. Bolstered by the court's decision, residents of Aztec rode into Junction City one night and retrieved the records. They remain in Aztec to this day.

At the Aztec Museum Pioneer Village you can view the Hamlet cabin, named after its occupant who hosted a fatal Christmas-eve party in 1880. Secure with its place as the county seat, Aztec was incorporated in 1905. Like the whole of San Juan County, its prosperity has fluctuated with the oil and gas industry. Aztec (1990 pop. 5,479; elev. 5,650 feet) is thirteen miles northeast of Farmington on U.S. 550. [3, 51, 85, 192]

Aztec Ruins National Monument. Two ancient peoples occupied this site. The first group may have been Chaco; at least they had a culture similar to the Chaco and arrived there during the eleventh century. They built the original pueblo but apparently abandoned the place before the second occupants, who were more culturally akin to the cliff dwellers of Mesa Verde, started using it a few decades later. In less than a century they also left. Why the groups came and went is unclear—changing weather patterns and availability of food both may have played a part.

Early settlers named the place from the false idea that the ruins had something to do with the ancient Aztec civilization of Central America.

The monument is at Aztec on the northwest side of the Animas River via Ruins Road. [161]

badlands. Dry lands eroded into strange, sometimes grotesque shapes. Their terrain, aridity, and inorganic soils render them valueless for farming or grazing. Few plants and animals survive in their forbidding habitat. The Angel Peak National Recreation Site and the Bisti and De-Na-Zin Wildernesses offer examples of such formations.

Baker, Charles (1820–?). This early basin prospector's reputation has reached legendary proportions, but surprisingly little is known about him. We don't know how he got the often-used title of "Captain," and some historians question even his given name.

Backed by Denver investors, he first came to the San Juans from upper Arkansas River diggings in 1860. He prospected through the Gunnison River country, then proceeded up the headwaters of the Lake Fork of the Gunnison and across what is today called Cinnamon Pass, with the park on the upper Animas River as his destination—the park that today bears his name. Among his likely companions were two others whose names became implanted in the area: W. H. Cunningham (Cunningham Gulch) and George W. Howard (Howardsville).

After prospecting during the summer of 1860, Baker chose to resupply his party not through passage to Denver, but from New Mexico. His party and others, realizing the difficulty of wintering in the San Juans, took refuge at a lower elevation in the Animas Valley. They chose a site on the east side of the river near present-day Rockwood. Here the adventurers rejuvenated their bodies in the hot springs on the west side of the river and drank from the cold springs on the east.

This settlement became the original Animas City. It lived for just a year.

Baker predicted that the untapped mineral wealth of the mountains and the agricultural potential of the valleys would attract 25,000 people

After ancient seas receded from the basin's lower reaches, wind and water eroded their sediment into badlands. Many feature unique formations, like this one in the Bisti Wilderness. Courtesy Farmington Museum, Catholic Charities Collection, 1992.8.1.

to the region within a year. (His prediction ignored the fact that, without the Ute's permission, the immigrants would be trespassing.) Baker's forecast was premature. In the spring of 1862 the sluice boxes remained unattended. Whether frightened by the Utes or attracted by the Civil War, Baker and his contemporaries chose not to pursue the mountains' hidden wealth. The rush would come, but later.

Baker's reputation as a pioneer is well deserved, for he not only prospected the hills. By using New Mexico for supply and support, he helped join the mountainous northern section of the basin with the older, Hispanic part to the south, thus helping fuse the San Juan Basin. [96]

Baker's Bridge. The log structure that spanned the Animas River's stark cliffs at the site of the original Animas City was named for Charles Baker, who started the settlement in 1860–1861. Future bridge builders carried the name forward. The modern steel and concrete bridge twelve miles north of Durango on La Plata 250, a few hundred yards east of U.S. 550, is south of Baker's original structure. [96]

Baker's Park. A small plain amid the precipitous San Juan Mountains. In what was a grassy meadow before the Anglo-American rush for gold and silver in the nineteenth century, Cement Creek and Mineral Creek join the Animas River before the larger stream leaves the park to flow south. Named after Charles Baker, an early prospector, it provides the setting for present-day Silverton. [96]

barbed wire. Contrary to the myth of the open range, this new technology turned out to be the cattlemen's best friend. In the 1880s they bought tons of it to defend "their" grass. Before its invention in 1874 by an Illinois farmer named Joseph Glidden, cattle raisers had fought fencing laws. But barbed wire proved to be a wise investment. It not only kept out the neighbor's livestock, it prevented cattle from drifting and thereby cut roundup costs. It also kept prize bulls from wasting their semen on competitors' cows. [89, 223]

Barker Arroyo. A gulch that runs south into the La Plata River ten miles upstream from the river's junction with the San Juan. It was named after a cowboy murdered there in a vendetta. In January 1881 members of the Farmington Stockmen's Protective Association, a vigilante bunch, had shot and killed Ike Stockton's brother, Port. Among those who took part in the murder were Tom Nance and Aaron Barker. Ike and his companion, Dyson Eskridge, sought revenge and caught up with the two vigilantes in the arroyo. Stockton chose Barker as his target and fired. Barker

Silverton and Baker's Park, ca. 1912. In 1860 Charles Baker led a prospecting party to this valley in the San Juan Mountains. After the Utes were forced to give up the region in 1873, mining flourished and the area prospered. Courtesy Fort Lewis College, Center of Southwest Studies.

fell dead from his horse and the gulch was forever christened Barker Arroyo. [85]

Battle Rock. According to legend, the name of this peculiar formation in McElmo Canyon got its name during a war between Native American tribes. A defeated group took refuge on its west slope, but the victorious tribe continued its pursuit and drove the vanquished over the precipice of the rock's east-facing cliff. [50]

Bayfield, La Plata County, Colorado. This town grew up as an agricultural center serving the Florida Mesa community and the Pine River valley. John Taylor, a former Kentucky slave, declared himself to be the "first white man in Pine River." (Although he was black, Taylor used the term "white man" to indicate he was not a Native American.)

A community with its own history, Bayfield is fast becoming a worker's suburb of Durango. If you are alert, you may catch the Bayfield parade in the movie *When the Legends Die* (1972, Fox-Rank/Sagaponack).

Bayfield (1990 pop. 1,090; elev. 6,900 feet) is eighteen miles east of Durango on U.S. 160. [167]

bear. The only bears known to live in the San Juan country are black bears; they come in several colors. Although they prefer the forest, black bears are most likely to be found where they can satisfy their omnivorous diet—in berry patches, creek bottoms with insects, or places with a lot of small mammals. If you leave food about in the woods, bears may be there too.

Grizzly bears once abounded in the basin and throughout the West. Early nineteenth-century trapper George C. Yount reported that it was not unusual for him to kill five or six of the fifty or sixty he might see in one day.

A hunting guide killed the last known grizzly in the San Juan Basin in 1979 as it attacked him near the headwaters of the Navajo River, twenty miles east of Pagosa Springs. Of the several reports of grizzlies in that vicinity since then, none has been confirmed. Even so, the Forest Service gives tips on venturing into grizzly country.

Black bears are protected as a game animal throughout the basin. If a grizzly is out there, it is protected by the Endangered Species Act. [35, 137, 178]

beaver. More than any other animal, these furbearers opened the mountains of the basin to European exploration. Trappers Ezekiel Williams and Robert McKnight worked in the San Juans as early as 1811. William Wolfskill and Ewing Young followed a couple of decades later.

And until silk hats replaced beaver pelts in the 1840s, mountain men crisscrossed the continent trapping beavers. How close these critters came to extinction no one knows, but biologists agree that a whim of fashion may have saved them.

These rodents, which can weigh up to fifty-five pounds, show their presence with dams, lodges, bank dens, and gnawed stumps. They maneuver underwater by using their webbed feet and broad, flat tails. Among the mammals, only humans have a greater influence on their surroundings. When beavers leave or are taken away, the ecosystem changes dramatically. That is why they are regarded as a riparian "keystone species." [35, 96, 135]

Beklabito, San Juan County, New Mexico. Trader Billy Hunter started a trading post here around the start of the twentieth century. The Foutz family later included the post in their string of twenty.

The post is nineteen miles west of Shiprock on U.S. 64. [87]

Big Bend, Montezuma County, Colorado. Settled in the 1880s during the area's cattle and homesteading era, Big Bend was so named because it was on the Dolores River where the stream (now submerged by McPhee Reservoir) turns abruptly northwest.

When the Rio Grande Southern Railroad came up the Dolores River valley in the early 1890s, it missed Big Bend and followed a route that crossed the river two miles upstream. Promoters there started the community of Dolores, and the settlers and merchants of Big Bend moved to the new community. [50]

bighorn sheep. Until the mid-1800s a million of these ovines may have ranged the West at various elevations. Also called Rocky Mountain sheep, most are found at higher elevations in the mountains of Utah and Colorado, some in the basin's San Juans.

The rams, which may stand forty inches at the shoulder and weigh up to 350 pounds, are known for their way of butting heads in a battle for the ewes. Their curling horns can attain a length of fifty inches. Bighorns like to feed in open grassy meadows next to rocky cliffs where they can watch out for predators and escape quickly.

The keys to the bighorn's survival are population control—both sheep and human—and better habitat. Regulated hunting can manage animal population; improving the habitat may require the restriction of other human activities. [145, 217]

Bisti Wilderness. A bizarre cluster of clay and rock formations near Farmington. When the water receded from an inland sea millions of years ago to change this part of the San Juan Basin from coastal swamp to inland floodplain, it left lush tropical plant life to rot. Nature has eroded the resulting mix of coal seams, shale, and sandstone into grotesque formations. Weathered sandstone forms spires and hoodoos (sculpted rock) throughout the wilderness. High temperatures have baked and oxidized coals to produce red hills.

Congress designated this seven-square-mile tract as wilderness in the San Juan Basin Wilderness Protection Act of 1984. In the Navajo language, Bisti (pronounced *bis-ta-hi*), means "badlands." Nonnatives commonly pronounce it *bis-tie*.

Administered by the Bureau of Land Management, the area is thirty miles south of Farmington on New Mexico 371 with access provided by Navajo 7000. [148, 171]

blacklist rule. In the frontier West, this dictum of cattlemen's associations' by-laws required that members report to the association any employee who was fired for dishonesty or cattle theft. The culprit's name appeared in the next association bulletin, and he was thereafter banned from employment by organization members. [53]

Blanco, San Juan County, New Mexico. This community started in 1901. Like many outlying settlements in the New Mexico desert, it served as a trading post. One of its early occupants was Frank Townsend, well known for the joviality that helped get him elected county commissioner. His business card revealed his sense of humor: "FRANK TOWNSEND, INDIAN TRADER, DEALER IN SORE-BACK HORSES, SCABBY WOOL AND SANDY SHEEP."

Blanco is nine miles east of Bloomfield on U.S. 64. [3, 85]

Blanco Trading Post. Tabby Brimhall started this trading post (not to be confused with the town of Blanco), probably in the 1920s, twenty-seven miles south of Bloomfield on New Mexico 44. [87]

Bloomfield, San Juan County, New Mexico. Colonizer John Bloomfield got $10,000 from the Mormon church in the 1880s to help pioneers build an irrigation system for the Porter and Hammond communities. Porter was renamed to pay tribute to him and to encourage him to stay around. He didn't, preferring instead to colonize further in Utah, Arizona, and other parts of New Mexico.

Here the Reverend Hugh Griffin first settled after coming to the region from Texas, even while the local cowboys did their best to encourage him to leave. One night Griffin was reading his Bible by the fireplace when some of the local hooligans entered the room and shot off their pistols at the fire. Griffin refused to react. By such calm demeanor he was able to earn the local ruffians' respect and stay around long enough to convert such miscreants as Farmington vigilante Alf Graves. Bloomfield (1990 pop. 5,214) is thirteen miles east of Farmington on U.S. 64. [3, 85]

blue spruce. In the East this splendid evergreen adorns more artificial landscapes than any other western conifer, and its blue shade brings a special bloom at Christmas. In the West it marches nobly up the slopes of the Rockies to an elevation of 11,000 feet. As its lower boughs die off from forest competition, the spruce's tops take on a sky-piercing quality. Also called the silver or Colorado spruce, it is Colorado's state tree. Its usual color is a somber green, but its new growth exudes an azure cast. [81, 233, 237]

Bodo State Wildlife Area. This 7,549-acre tract lies south of Durango at elevations of between 6,800 and 7,400 feet. Its mountain browse of oak brush and piñon, mingled with ponderosa pine, provide forage for deer and elk. Bears and mountain lions frequent the area, as do rabbits, blue grouse, doves, and band-tailed pigeons. Golden eagles and other raptors nest in nearby cliffs. If the Animas–La Plata Project is built, a reservoir will inundate part of the range. The area has significant ancestral Puebloan archaeological sites.

Bodo is on La Plata 211 a few hundred yards west of U.S. 550/160 and a mile south of Durango. [250]

Bondad, La Plata County, Colorado. The opening of the Ute Strip in 1899 stimulated settlement on this part of Five Mile Mesa in southern La Plata County, but not enough to form a town. The lure of cheap land did generate animosity, however. As land seekers made a run from Twin Crossings and drove stakes on their spots of choice, claims overlapped. Those with money bought off contenders. Others became discouraged and left.

Bondad is a couple of miles north of the Colorado–New Mexico line on U.S. 550. [191]

Bosque Redondo. This gentle bowl, termed by the Spanish a "round forest" of cottonwoods, served as General James H. Carleton's containment camp for Native Americans of the Southwest, especially Navajos driven from the San Juan Basin. The camp was located near Fort Sumner, New Mexico, some 300 miles southeast of Farmington.

After ordering the Mescalero Apaches confined there in 1863, Carleton saw it also as the place to imprison the Navajos. He forced these natives to surrender through his scorched-earth policy, executed primarily by Colonel Kit Carson, and then embarked on the second stage of his scheme to subjugate the Navajo people—a plan euphemistically called "colonization."

By 1864 over 8,000 Navajos had given up and were forced to take the "long walk" to the 1,600 square miles of Bosque Redondo. Even with irrigation the region was pitifully inadequate for agriculture. Floods, droughts, hail, and insects took their toll on the crops. Sheep and goats suffered for lack of forage; some were stolen by marauding Kiowas and Comanches. Meager government rations of weevil-infested flour and rancid bacon barely held off starvation. Pneumonia, measles, and other diseases attacked the weakened prisoners. Fights broke out between the Navajos and their traditional enemies, the Apaches. Fraud among agency administrators made an abject situation even worse.

Not until the army reorganized the Department of New Mexico and transferred Carleton to another position did higher officials give Bosque Redondo an objective look. The absolute poverty they found led them to concede failure and let the Navajo people reverse their "long walk" and return to their (now diminished) homeland. [80, 82, 123]

breaks. The rough, gullied terrain where mountain foothills give way to the flatlands. New Mexico 574 goes through such land west of Aztec.

Breen, La Plata County, Colorado. A community named after Thomas H. Breen, an early superintendent of the Fort Lewis Indian School. Located nearby, the school was established in 1892 and was headed by Breen from 1884 to 1903. In 1905 he was granted a 240-acre homestead where the Breen store still stands. In pioneer days this farming community had a post office.

Breen (elev. 7,550 feet) is southwest of Durango, eleven miles via U.S. 160 and seven miles on Colorado 140. [49]

Brunot Agreement. An arrangement named after U.S. Indian commissioner Felix Brunot whereby the Utes gave up for Anglo settlement a section of the San Juan Mountains that had been invaded by white miners. An 1868 treaty had reserved for the Utes most of that portion of Colorado sloping west from the Continental Divide. Miners ignored the treaty, prompting the U.S. government to negotiate with the Utes for the land occupied by the miners.

The ensuing accord, also known as the San Juan Cession, became law with President Ulysses S. Grant's signature in 1874. It cut a rectangle of more than 6,000 square miles out of the 1868 reservation. [74, 86]

Buckskin Charlie. See **Sapiah.**

buffalo (bison). Although some historians include the northeastern part of the San Juan Basin in the original extent of the bovine's range, this denizen of the plains played little direct part in the basin's history. To harvest the animal or barter with plains tribes for hides, Utes and Apaches had to go over the Continental Divide to the Rocky Mountains' eastern foothills. [204]

Buffalo Soldiers. When they saw the troops coming up the Animas Valley in October 1879 to allay their Ute-inspired anxieties, the citizens of Animas City couldn't have cared less about the horse soldiers' color. At that time African Americans comprised a fifth of the army's cavalry.

In support of companies from the Twenty-second and Fifteenth Infantries were some of the Ninth Cavalry's "black white men." When they served under the flag of the U.S. Army, Native Americans called them "Buffalo Soldiers."

After the Civil War, Congress authorized four black battalions, the Twenty-fourth and Twenty-fifth Infantries and the Ninth and Tenth Cavalries. Of the two cavalry units, the Tenth became the better known, largely because Frederick Remington and others reported and illustrated its exploits. (In his office chairman of the Joint Chiefs of Staff Colin L. Powell hung a Don Stivers print, *Tracking Victoria,* depicting soldiers of the Tenth Cavalry pursuing an Apache warrior.) But for the fearful settlers of the San Juan Basin, the Ninth Cavalry, with mobility to match the mounted Utes, was their savior. After riding from Fort Lewis, then garrisoned at Pagosa Springs, companies of the Ninth bivouacked at Animas City. A few years later they would play another native-subjugating role for the basin, relocating the Jicarilla Apaches to their reservation in Rio Arriba County, New Mexico.

During the Indian Wars, 12,000 African Americans served, and 12 received Congressional Medals of Honor for bravery in combat. "On horseback, in his coat of blue, eagles on his buttons, crossed sabers on his canteen, rifle in hand, pistol on his hip, brave, iron-willed, [the Buffalo Soldier was] every bit the soldier that his white brother was." So said Chairman Powell as he helped dedicate a monument to the Buffalo Soldiers at Fort Leavenworth, Kansas, in 1992—well over a century after they helped make the basin safe for whites. [103, 197, 222, 234]

Bureau of Indian Affairs. Reflecting the status of the Native American tribes in the eyes of expansionist whites, Congress set up this agency as part of the U.S. War Department in 1824. In 1849 it moved to the Department of the Interior. The agency serves as the trustee for tribal land and money held in trust for tribes by the United States. It is concerned with about 500 tribes, including Alaska native villages. Its mission is to help the country's two million Native American people, most of whom live on 300 reservations, develop their human and natural resources.

In the basin, the bureau is concerned with the Jicarilla Apache, Navajo, Southern Ute, and Ute Mountain Ute tribes. [220, 230]

Bureau of Land Management. The General Land Office and the Grazing Service combined to form this bureau in 1946. This agency, within the U.S. Department of the Interior, administers an area the size of Texas and California: the 422,000 square miles of public-domain land not assigned to other agencies for national parks, forests, or other specific uses. It also manages the government's mineral rights under an additional 470,000 square miles of the country. The bureau regulates the use of all resources on land for which it is responsible, including timber, minerals, oil, rangeland vegetation, wild and scenic rivers, and wild horses and burros.

In addition to thousands of acres within the basin, the bureau manages the Anasazi Heritage Center, Angel Peak National Recreation Site, De-Na-Zin Wilderness, and Bisti Wilderness. [230]

Bureau of Reclamation. Within the U.S. Department of the Interior, this agency manages, develops, and protects water and related resources on the arid lands of the West's seventeen contiguous states. (Back east, the Army Corps of Engineers does this sort of thing.) It was first authorized in 1902 as a part of the Geological Survey's Reclamation Service.

Among its 322 storage dams, 14,490 miles of canals, 174 pumping stations, and fifty hydroelectric plants throughout the West are seven constructed San Juan Basin projects: Dolores, which includes McPhee Reservoir on the Dolores River near Dolores; Florida, which forms Lemon Reservoir on the Florida River fourteen miles northeast of Durango; Hammond, a diversion dam and irrigation water delivery system along the southern bank of the San Juan River opposite Blanco, Bloomfield, and Farmington; Mancos, where the Mancos Reservoir (Jackson Gulch Reservoir) originates five miles north of Mancos; Navajo Indian Irrigation, south of Farmington; Navajo Storage Unit, located on the Colorado–New Mexico line near Arboles; and Pine River, which forms Vallecito Reservoir on Los Piños River (Pine River) eighteen miles northeast of Durango.

The proposed Animas–La Plata Project near Durango is also a bureau undertaking. [230, 265]

Burnham, San Juan County, New Mexico. The first trading post at this place on the Navajo Reservation was built by Roy Burnham about

Animas City burro corral, ca. 1880. Because it could find its own food and carry several hundred pounds, the hardy burro was the prospector's pack animal of choice. Courtesy La Plata County Historical Society.

1927. His father, Mormon bishop Luther C. Burnham, came to the basin in 1878 with his family (or families) from St. Johns, Arizona, with the encouragement of Brigham Young Jr.

Burnham is twenty-nine miles south of Shiprock on U.S. 666 and then twelve miles east on Navajo 5. [85; 87]

burro. The Spanish name for donkey. In the West this equine proved indispensable for prospecting; some carried up to 300 pounds of provisions. Due to their wanderlust and stubborn streak, burros are the subject of many entertaining stories, such as the one about the prospector who, having spent fifty years searching for gold, admitted he spent thirty of them looking for his burros. [223]

Cahone, Dolores County, Colorado. This community started when homesteaders came to the Dolores River valley in the early 1900s, but it didn't get a post office until 1917.

Cahone (elev. 6,115 feet) is twenty-six miles north of Cortez on U.S. 666. [50]

canyon. A large, narrow gorge usually formed by stream erosion, such as Largo Canyon, which stretches south from the San Juan River upstream of Farmington. A small canyon may be called an arroyo. The largest of these natural features in the West is of course the Grand Canyon.

Capote. A Ute band whose hunting territory was primarily east of the Continental Divide on both sides of the Colorado–New Mexico border but stretched west into the San Juan Basin as far as the Animas River. Many of their descendants are members of the Southern Ute Tribe. [119]

Carbon Junction, La Plata County, Colorado. A railroad camp where the Farmington branch of the Denver and Rio Grande Railway took off from the line that continued east. It was near the present-day intersection of U.S. 160 and U.S. 550 about three miles south of Durango.

Carson, Kit (1809–1868). Trapper, scout, soldier, and Indian agent whose military career and treatment of Native Americans profoundly affected the San Juan Basin. As an army colonel he drove the Navajos out of the basin. As a civilian he went to Washington, D.C., on behalf of the Utes.

Given the name Christopher Houston Carson, he grew up in Missouri, the descendant of Kentucky frontier stock. At age seventeen, he fled an apprenticeship as a saddle maker to join a wagon train bound for

Santa Fe. Trapping beaver with Ewing Young, Carson became familiar with frontier trails and learned the skills for survival in the wilderness amid hostile natives. In 1841, with two children to support and beaver becoming scarce, Carson went to work as a hunter at Bent's Fort on the Arkansas River in present-day Colorado.

On a Missouri River boat trip the next year, he met John Charles Fremont and became the explorer's expedition leader. Carson's guidance led Fremont to become known as the "pathfinder." With his knowledge of the West, Carson became a valued army courier during the Mexican-American War as he carried Fremont's dispatches from California to Washington, D.C., passing through the basin on the Old Spanish Trail. As an army scout he took part in an engagement that helped establish California's Bear Flag Republic.

While an Indian agent operating out of Taos, New Mexico, he accompanied several expeditions against the Apaches and Utes but retained a sympathetic understanding of their plight. As a colonel in the First New Mexico Volunteer Infantry, Carson saw action at the battle of Valverde, but his subjugation of Apaches and Navajos held more importance for the basin.

In 1862–1863, under the command of General James H. Carleton, Colonel Carson operated out of Fort Stanton in southeastern New Mexico to defeat the Mescalero Apaches. He followed that with a scorched-earth policy against the Navajos as he destroyed their crops, burned their villages, and defeated their warriors at Canyon de Chelly. In November 1864, while Colonel John M. Chivington was massacring Cheyennes at Sand Creek in southeastern Colorado, Carson used a force of 410 men and two howitzers to hold off hordes of Kiowas, Comanches, and other natives at Adobe Walls in the Texas Panhandle.

As a civilian, in 1868, he made his last trip to Washington—a mission on behalf of the Utes, demonstrating his sympathy for the sad predicament of those Native Americans.

In contrast to a larger-than-life legend, Carson was small and stoop-shouldered with a freckled face that belied his courage. And according to some historians, Christopher Houston Carson was always the reluctant warrior. [58, 221]

Carson National Forest. Of this 2,354-square-mile reserve, only 200 square miles (the Jicarilla Ranger District) lie on the west slope of the Continental Divide and in the basin. The district's eastern and southern borders run against the Jicarilla Apache Reservation while its northern boundary touches the Colorado–New Mexico line. This isolated section covers high mesas of ponderosa pine and piñon-juniper forests. A herd of wild horses lives there. [155]

C

Cascade, La Plata County, Colorado. In 1880 this stage station offered the traveler a hotel and post office housed in a cabin at the stage road's summit on Cascade Hill overlooking the Animas River canyon. For those needing horse or mule feed, the proprietor offered hay and hand-harvested native grasses. The way station was busier than many in the region, for it linked Silverton to the north with Rico to the west and Animas City to the south.

The Cascade station was about twenty miles south of Silverton on a stage road that climbed up from the Animas Canyon. Forest Service 591 off U.S. 550 twenty-five miles north of Durango traverses the vicinity. [248]

Cather, Willa Sibert (1873–1947). While most of her earlier works depicted the lives of immigrant families on the Great Plains, this author's travels in the San Juan Basin and the Southwest inspired the settings for many of her later novels.

When she was ten, her family moved from Virginia to Red Cloud, Nebraska. After receiving her college education at the University of Nebraska, she was a journalist and a teacher in Pittsburgh before she became editor of *McClure's Magazine* in New York City. She left the magazine in 1913 to devote herself solely to writing.

In 1915 she traveled through Denver to Durango and rode the Rio Grande Southern Railroad to Mancos, an experience she noted to be remarkably friendly. While pausing six days at Mancos (she had intended to stay only one), Cather let the suspense of a deferred excursion to Mesa Verde add to its drama. When she finally succumbed to its beckoning, she was awestruck by its setting and history. From her visit to the mesa came the setting for sections of *The Professor's House* (New York: Alfred A. Knopf, 1925) about the ancestral Puebloans.

Among the novels to come out of her Southwest experience was *Death Comes to the Archbishop* (New York: Alfred A. Knopf, 1927), based on a Roman Catholic bishop's experiences among New Mexico's Native Americans. Even before she witnessed Mesa Verde, Cather was enthralled by the pre-Columbian people of the Southwest, as evidenced by her haunting 1909 short story, "The Enchanted Bluff." [199, 217]

cattle. Western lore leads us to believe that the Texas longhorn was the only breed the nineteenth-century cowboy ever branded. The basin's early cattle experience helps dissolve that myth.

While it is true that the herding of these beasts out of Texas brought the largest migration of domestic animals in the world's history, the stampede ran headlong into controversy and resentment. The reason:

on many ranges other breeds were there first. Such was the case in the basin.

Early in the 1860s, Fernando James and Frank Wadsworth herded their Missouri shorthorns into the basin, though it was not their range of choice. They first took their cattle to southeastern Colorado, but the Utes didn't like that idea. So the cattlemen drove their stock over Raton Pass to Santa Fe and Taos, then north to the succulent grasses of the upper San Juan Basin.

The climate was satisfactory, water adequate. Forage in the form of virgin wheatgrass and other grasses was plentiful—growing high enough "to brush a horse's belly." Trees, mountains, and gulches furnished shelter from the storms of winter and the heat of summer. There not only did James and Wadsworth persuade the Native Americans to let them pasture the animals on the rich grasslands of what is now Montezuma County—they got the natives' help. The cattlemen placed small herds in the care of various Ute families, provided supplies and horses, and let them butcher animals for their own use. In exchange, the cattlemen got a share of each year's calf crop. As would the Anglo-Americans in future generations, the Utes drove their cattle to the high mountain meadows for summer grazing, then brought them down in the fall. This arrangement, which continued into the 1870s, provided the breeding stock for many of the basin's better shorthorn herds in later years.

Meanwhile, Charles Goodnight, James Sheek, and other Texans were taking advantage of the post–Civil War demand for cattle by herding their stringy, tick-infested longhorns into New Mexico. The wild Texas cattle had a resistance to the virulent tick fever, but the "American" stock had no protecting antibodies. When the better stock crossed a longhorn trail, they attracted ticks scattered by the Texas cattle and died in droves. And cows that managed to avoid the plague were impregnated by longhorn bulls.

Colorado's territorial legislature outlawed the entry of Texas cattle into its southern counties, but enforcement was impossible. Some years later the competing parties worked out an arrangement to winter the longhorns in central New Mexico and kill the ticks before trailing the wild stock farther north. This helped resolve the problem so the basin's cattle industry could prosper. Livestock ventures got another boost when the Denver and Rio Grande Railway arrived in 1881 and made Durango a livestock shipping point.

The basin's cattle business has outlived post–World War I deflation, the Great Depression, vacillating federal range policies, and fickle consumers. It flourished after World War II and has since swung with the economy and changing dietary preferences.

The rough mountains of the basin diverted the legendary Texas trails, so what longhorns did get there arrived by lesser routes. It has been over a century since herds wandered freely on the range. The Hereford and the Angus have joined the Missouri shorthorn on the pastures. The fragrant semis and traffic-stopping drives along the highways are proof of the cattle industry's endurance. [53, 54, 204, 208]

cattle brands. When a San Juan Basin cowboy sears an insignia on an animal's hide, he follows a practice dating back to the Egyptian pharaohs. The Spaniards adopted the idea and brought their branding irons to the New World.

The proliferation of cattle outfits after the Civil War brought confusion to cattle branding. Counties tried to loop the situation by registering brands, but with ranges and strays crossing county boundaries, duplications persisted. Not until the states took over and Colorado Livestock Inspection Board secretary W. C. Baker came up with an orderly method of brand classification did the system become manageable. When Colorado and New Mexico (and the other western states) adopted the Baker system, the basin's cattlemen—at least the honest ones—must have been grateful.

During the cattlemen's heyday in the 1870s and 1880s, H. W. Cox and his sons ran the largest outfit in the basin from their headquarters in the Animas Valley at the Colorado–New Mexico line. The COX brand adorned as many as 7,000 cattle. Second in size was G. W. Thompson's Two Cross enterprise along the La Plata River. Its emblem was a pair of interlocked crosses. The Carlisle Cattle Company, run by Ted, Howard, and Thomas (relationship unknown), dominated the range south of the San Juan River east of the Arizona line. They used the Bar-U brand.

Even though the propane burner has replaced the campfire and it's illegal to run cattle without a registered brand, keeping track of ownership is still a tedious job. The brand books of Colorado and New Mexico together have over 58,000 entries.

In addition to symbols on their hides, some cattle carry implanted microchips. Perhaps branding irons will soon be relegated to museums and brand books to archives. Then, in his scabbard, the basin cowboy will carry a scanner with a light-emitting diode. [53, 208, 210, 226]

cattle drives. When H. W. Cox, with his sons and son-in-law Alf Graves, brought 4,500 cattle to the basin from Stephenville, Texas, in the 1880s, they likely used the system familiar to Texas drovers. A typical trail crew was made up of a trail boss, fifteen or more cowhands, a cook, and a horse wrangler. The crew would put a lead steer at the head of the herd—perhaps one they had used before and knew to be reliable. The cattle stretched behind the decoy animal, somewhat bunching up at the rear of the herd.

Three cowboys rode on either side of the migrating cattle. Those riding close to the lead steer were the "point" men; behind them, near the middle, were the "swing" men. Near the rear the "flank" men urged the herd along with the rest of the trail crew; the "drag" men were strung in a semicircle behind to catch any strays.

The biggest challenge to the drovers often came at night when a bolt of lightning or some mysterious occurrence would spook the herd to stampede. When this happened, it was the job of the trail hands to gallop along the herd's flank and turn the lead animals until the whole bunch was milling in a circle, a process that might continue all night.

Daily, the herd moved no more than twelve to sixteen miles, so the trek from north-central Texas to the basin could have taken up to three months. If the drovers wintered the longhorns in New Mexico to rid them of fever-infesting ticks, the journey took much longer. [194, 223]

See also **cattle trails**.

cattlemen-sheepherders wars. Even Hollywood couldn't exaggerate the antagonism between these two breeds. And until the West matured and justice prevailed, the cattlemen usually won.

When Port Stockton, riding with a horse-thief posse past Cedar Hill, happened upon a Hispanic sheepherder, he simply shot the herder dead and the posse rode on. Port was never arrested for the murder, possibly because the rest of the bunch considered the incident too trivial to report. Ethnic prejudice may have come into play or, as was common, the sheriff perhaps sympathized with cattlemen and simply neglected to act. In the winter of 1885–1886 two Durango cowboys riding for the Carlisle outfit shot and killed Ricardo Jacques, a New Mexico sheepherder. The New Mexico attorney general brought them to trial, but the proceedings were a farce and there was no conviction.

Short of murdering sheepherders, cattlemen often just got rid of their flocks. During the era of conflict, from about 1870 until 1920, cattlemen shot, clubbed, poisoned, knifed, dynamited, or otherwise disposed of more than 53,000 sheep—14,000 in Colorado but fewer than 1,000 in New Mexico. (Hispanic New Mexicans had always preferred sheep over cattle, so sheepherders had the upper hand.)

The quest of various groups for a slice of the shrinking public lands aggravated the cultural animosity between cattlemen and sheepherders. As homestead settlement reduced the open range, competition became fierce. Much of the forest land that had been grazed by nomadic sheep flocks was withdrawn from the public domain in 1891, forcing sheepherders to invade cattle country. As throughout most of the West, range wars broke out in the basin. Whether the strife between cattlemen and sheepherders was worse than that between cattlemen and homesteaders is a matter for conjecture. It was apparently more violent. In either case, it formed one of the greatest, and most shameful, dramas of the basin and the West. [43, 54, 85, 98]

cattle trails. Civil War veterans returning home found perhaps six million feral longhorns in the broad arrowhead of South Texas, a commodity waiting to be delivered to the markets of a meat-hungry nation. The routes that splayed out from that triangle were so numerous that "cattle trails" and "Texas cattle trails" are synonymous.

The Shawnee Trail, through present-day eastern Oklahoma to Baxter Springs, Kansas, was the first to gain notoriety. As the railheads moved west, so did the trails to meet them. The Chisholm Trail went to Wichita, Newton, and Abilene. Other trails headed west to California. Unlike the drovers who headed for the railheads, Charles Goodnight saw opportunities at the forts, Indian agencies, and mining camps farther west. He herded his cattle up the Pecos River to Fort Sumner, New Mexico, then to Santa Fe.

No famous trails ran through the San Juan Basin; it lay west of the routes to the railheads, and its terrain forced the California trails to stay south. Goodnight's route through Santa Fe came closest to the basin. Its longhorns came from there.

Goodnight and Jim Loving used their trail to drive herds north through eastern Colorado to Wyoming, Montana, and on into Canada. During the longhorn's prime years, 1866–1886, ten million of them trampled the trails to leave an indelible mark on the saga of the West. [108, 204, 208, 226]

Cedar Hill, San Juan County, New Mexico. In the 1890s, on the mesa above this community, a resourceful settler captured rain to irrigate his vineyard. The mesa was also a scene for the popular sport of horse racing. Cedar Hill is three miles south of the Colorado–New Mexico line on U.S. 550. [85]

Chaco. In the eleventh century these ancestral Puebloans were the economic masters of the San Juan Basin. From their cultural center in the southern part of the basin, they designed and built 400 miles of roads—not just worn paths—to remote villages.

With imported turquoise they made jewelry to exchange for pottery, seashells, and copper bells from other ancient people living as far away as present-day Mexico. Using their own special techniques, they constructed four-story building complexes. In and around their largest

community (in what is now the Chaco Culture National Historic Park),
as many as 5,000 people may have lived in some 400 settlements. Artifacts and structural remains tell us they first built there in the ninth century and reached their apex of economic development in the eleventh,
then drifted away in the twelfth. Archaeologists speculate that drought
and diminishing resources forced them to leave. Their descendants still
live among the Pueblo people. [142, 161]

Chaco Canyon. The main defile within the Chaco Culture National Historic Park. This place-name commonly serves as an informal reference
to the park.

Chaco Culture National Historic Park. Designated as a World Heritage Site, this location joins a select list of protected areas around the
world. Within its thirty-three square miles are the largest excavated prehistoric ruins in North America, those of a people whose culture centered here during the tenth through twelfth centuries—the Chaco.
   A route to the park, beginning with San Juan 7900, leaves New Mexico 44 three miles south of Nageezi. The park's visitor center is some
twenty-six miles from New Mexico 44 over mostly unpaved roads. [142]

chaparral. A thicket that may be composed of Gambel oak, brambles, or
other brush. The term comes from *chaparro,* Spanish for "evergreen
oak." [223]

Chimney Rock Archaeological Area. The most remote site of the
Chaco Canyon–centered ancestral Puebloans. Perched on a high mesa
overlooking the Piedra River, the area has sixteen excavation sites and
several Chacoan structures. Chimney Rock Pinnacle dominates the site.
(A peak with the same name in Nebraska gained fame as a landmark for
nineteenth-century wagon trains.)
   For reasons unclear, pre-Columbians sent forth squads to establish
roads and outposts. Scientists theorize that the site's natural formations
may have appealed to a religious need: here they could watch the moon
rise between pinnacles at the end of the "lunar standstill," a phenome-
non that occurs every eighteen years. They may also have used it as a
trading site. And since they employed large timbers for building at their
central Chaco Canyon site (where no large trees grew), Chimney Rock
may have been their source of supply.
   Chimney Rock is in the San Juan National Forest seventeen miles
west of Pagosa Springs on U.S. 160 and three miles south on Colorado
151. [156, 223]

Chimney Rock. This pinnacle near Pagosa Springs marks the farthest outpost of the Chacoan culture and commerce that fanned out from Chaco Canyon during the eleventh century. Courtesy San Juan National Forest.

chipmunk. The Colorado Plateau is home to the Colorado chipmunk, sometimes referred to locally as the Hopi chipmunk. In the basin you are more likely to see the Least chipmunk. Both are among several species found in the West.

Chipmunks are energetic little mammals that differ from other striped squirrels in having stripes on their faces. They like seeds, berries, insects, flowers, and picnic vittles. Since they do not store fat for deep hibernation, they use their cheek pouches to cache food in burrows for winter sustenance. [135]

Chromo, Archuleta County, Colorado. This was likely a timber camp in the late nineteenth century.

Chromo (elev. 7,500 feet) is twenty-four miles south of Pagosa Springs on U.S. 84.

Chuska Mountains. The principal range of these mountains lies along the Arizona–New Mexico border on the Navajo Reservation. The southern regions of this range were not surveyed until 1917, years after many of the basin's rugged San Juans had already been mapped.

The Navajos used this chain of mesas, cones, and gorges in their struggles with Spaniards, Mexicans, and Anglo invaders. Washington Pass provides a break through the range and was the scene of many skirmishes. The range's name is a corruption of *shashgai,* Navajo for "white spruce."

After the Navajos were permitted to leave Bosque Redondo and return to their reservation in 1868, the mountains proved a barrier to east-west commerce. As a result, a chain of trading posts grew along their eastern foothills. [243]

Cinnamon Pass. Up the Animas River from Silverton, before you get to Animas Forks, Cinnamon Creek joins the river from the east. Cinnamon Pass is near the headwaters of the creek. Over the pass lie the origins of the Lake Fork of the Gunnison.

Named for the cinnamon color in the mountain to the south, the pass saw Charles Baker and other prospectors come through to enter the San Juan Basin in 1860. Despite its high elevation, the route over the pass was an all-year mail route when the mines prospered in the 1870s. It is now on the Alpine Loop Byway.

The road over Cinnamon Pass (elev. 12,620 feet) leaves Colorado 110's extension near Animas Forks. [209, 219]

Civil War. Even though the "War of the Rebellion," as it was officially known, saw no armed conflicts in the San Juan Basin, the people of

C

Colorado and New Mexico played an important role in preserving the Union.

Both the Union and the Confederacy needed the West. The Confederacy sought its gold and silver to finance the war and coveted California's blockade-free ports (though you might raise an eyebrow at the idea of transporting war materials across the continent by pack mule and wagon train). The mission of securing New Mexico Territory for the South fell to the Texans. That accomplished, they would then march to Denver, secure the Colorado gold fields, and continue on to San Francisco. The first Colorado regiment of volunteers—nicknamed the "Pike's Peakers"—had to stop them.

After the Texans fought several successful engagements, Confederate Brigadier General Henry Hopkins Sibley, with the inflated title of Commander of the Army of New Mexico, found himself in Santa Fe relishing his victories but short of supplies. Colonel E. R. S. Canby, commanding the Pike's Peakers, had expected the Confederates to overextend themselves. Sibley was in a logistical trap.

In March 1862 Union Major John M. Chivington's advance troops of Coloradans marched the Santa Fe Trail through Glorieta Pass. In the southern Sangre de Cristo Mountains, 120 miles southeast of Farmington, he engaged the Rebel forces at Apache Canyon. A series of skirmishes proved indecisive until Chivington's troops made a risky flanking movement over mountainous terrain and destroyed the Confederate's supply camp. Following a last-gasp artillery duel in Albuquerque, the Rebels retreated down the Rio Grande, urged on by sporadic cannon shots from the Coloradans.

The Battle of Glorieta Pass quashed the Confederacy's hope of conquering the West. [77, 78]

Coal Bank Pass. This divide over Coal Bank Hill separates the watersheds of Coal Creek to the north and Mill Creek to the south. It offers views of Engineer Mountain two miles to its west.

Coal Bank Pass (elev. 10,640 feet) is thirty-five miles north of Durango on U.S. 550. [219]

Code of the West. "When you call me that, *smile!*" This threat by the Virginian in Owen Wister's novel of the same name (New York, MacMillan, 1902) is the epitome of the Code of the West. Here are the code's basic precepts.

No duty to retreat: The English common law requiring flight to avoid deadly conflict simply could not survive on a frontier with no assurance that authorities would deal with an offending opponent.

A man's survival seemed to depend entirely on his willingness to take on the aggressor.

Personal self-redress: To rely on others, even the legal authorities, to settle a grievance was considered to be "unmanly." Self-redress was not just a right; in the West it became a cultural duty.

The homestead ethic: This included the right to have and to hold a family-size farm, the homestead, and the right to enjoy it free from fear.

Individual enterprise: Those on the edge of the frontier—the trappers, miners, and homesteaders—concerned themselves with commerce only as a means to survival. Later came the entrepreneurs: mining companies, railroads, and cattle corporations. This gave rise to conflicts between the "little man" and the "incorporators." Examples of this antagonism abound in the reality of western history as well as in lore.

A high value on courage: Without courage, a man could not uphold the other tenants of the code, notably the ethics of not retreating and of redressing personal wrongs. But for many, this value became not just an end but a means. Thus, a misguided renegade with no legitimate reason to show his lack of fear might look for a handy opportunity ("When you call me that, *smile!*").

Other parts of the frontier code supported the basic tenants—don't be a quitter; be loyal to your "brand" or group; meet the obligations of friendship; play fair; don't use treachery; regard every quarrel as a private quarrel and don't interfere.

And pride, like courage, blanketed the code.

Events in the San Juan Basin gave credence to the Code of the West. As long as the notorious basin outlaw Ike Stockton met the obligations of friendship, loyalty, and fair play, he rode on the wrong side of the law with impunity. But when he turned to treachery and betrayed one of his own "brand," death soon followed. [20, 89, 121, 131]

Code Talkers. A World War II platoon of Navajos in the Pacific campaign who frustrated the Japanese by sending military messages in their native language. Japanese cryptographers could break a code, but they could never understand the Navajos. [82]

C

Coe, George Washington (1856-1941). Sidekick of Henry "Billy the Kid" McCarty, friend of Ike Stockton, and veteran of the Lincoln County War who brought his gunfighting experience to the basin.

The son of a Missouri Civil War veteran, Coe migrated to Fort Stanton, Lincoln County, New Mexico, in 1874 and went to work ranching for his cousin, Lou Coe. Then he leased his own spread. Meanwhile, conflicts between competing commercial interests approached all-out war.

In 1878 Coe was arrested on a phony charge by Sheriff William Brady and tortured in jail. This sealed his allegiance to the anti-Brady faction, the "Regulators." During the Lincoln County War's biggest shootout, Coe operated from Ike Stockton's saloon until he could move under cover of darkness to Alexander McSween's besieged headquarters. Then he manned a nearby warehouse to keep up the fight. When Billy the Kid and his allies broke out of McSween's burning house, Coe used the distraction to escape.

After the "possemen" shot McSween to death and the violence simmered down, Coe came with his family and other relatives to the basin. Here he took up farming and ranching near Farmington. When Ike Stockton followed, he gave Stockton temporary shelter.

In 1884 Coe left the basin. After sojourns to Missouri and Nebraska, he made his permanent home back in Lincoln County. There he secured a pardon from New Mexico governor Lew Wallace and homesteaded a place that became known as the Golden Glow Ranch. [37, 231]

Colorado. The Centennial State is home to 40 percent of the basin's 150,000 people; it joined the union in 1876. Its capital, Denver, is 332 miles from Durango whereas New Mexico's capital, Santa Fe, is more than a hundred miles closer to most of the basin's Colorado residents.

Even though the basin's Coloradans amount to less than 2 percent of Colorado's population (1990 census, 3,294,394) and its Colorado portion covers only about 6 percent of the state's area, the San Juan Basin's dramatic landscape and sensational past have molded a symbolic niche in the history of its mother state.

The state's symbols: flower, Rocky Mountain (blue) columbine; tree, Colorado blue spruce; bird, lark bunting; animal, bighorn sheep; gemstone, aquamarine; colors, blue and white; song, "Where the Columbines Grow"; fossil, Stegosaurus; insect, hairstreak butterfly. Motto: Nil Sine Numine, "Nothing Without Providence." [237]

Colorado–New Mexico border. An imaginary line through the middle of the San Juan Basin that was drawn through political skulduggery. In 1861 Congress split the vast region ceded by Mexico to the United States at the end of the Mexican-American War and formed two territories. To do so, it ran an east-west line along the 37th Parallel through the middle of the San Juan Basin to set the basin's northern part in Utah and its southern part in New Mexico. This arbitrary division began the split of a cultural whole.

For decades Spanish and Mexican cultural and political influence had prevailed. And while the Anglo prospectors labored to the San Juans over the Continental Divide from the north, once in the basin they and the settlers who followed them found it convenient to avoid the rigors of the divide and go south for supplies. New Mexican traders gladly brought their freight north to the new markets—as they still do.

The original Organic Act to set up the Colorado Territory as introduced on February 2, 1961, called for its southern boundary to go along the Arkansas River and trace a line well north of the basin, leaving out the Hispanic-settled regions on both slopes of the Continental Divide. This arrangement would have kept much of what is now southern Colorado within New Mexico and left open the possibility that the entire basin might become part of New Mexico. But Congress passed a different version only four days later. It dropped the new territory's south border to the old Utah–New Mexico territorial line—the 37th Parallel. And the revised bill contained unusual wording. It said the Colorado territorial legislature could pass no law "destructive of the rights of private property." Who changed the boundary to embrace a culturally different region and why?

The additional area encompassed some questionable Mexican land grants and, sure enough, New Mexican lobbyists had worked for the act's passage. The land grantees were removing their claims to a friendlier jurisdiction—one whose charter had an extraordinary protection for private property. Greed prevailed over territorial loyalty. [78, 117]

Colorado Plateau. The raised 140,000-square-mile terrain that includes the San Juan Basin. It spreads west to Nevada, north past Vernal in northern Utah, and south to Holbrook in central Arizona. To the east and northeast it stops at the Rocky Mountains. The plateau holds spectacular natural features. [236]

Colorado River. Had the Spaniards foreseen how their Anglo-American conquerors would squabble over this stream's water, they may have called it El Rio de Mucho Controversia.

All water falling on the San Juan Basin tries to get to the Colorado, for the basin lies within the river's 244,000-square-mile, seven-state drainage. The stream starts in northern Colorado on the west slope of the Continental Divide. What's left of its 1,400 miles later trickles into Mexico's Gulf of California.

Near the Four Corners area, the San Juan carries most of the basin's water into the larger stream—big by western standards but not heavy flowing. (It would take thirty-three Colorados to equal the roll of the Mississippi.)

The fights for the Colorado's water go far beyond the river's watershed. Hundreds of miles of aqueducts take its water to Denver, California's Imperial Valley, and the Southern California megalopolis. The drive for its water has forged international agreements, interstate compacts, Indian treaties, and a legacy of laws and court decisions destined to help water lawyers feed their families for generations.

Many arrangements, such as the Colorado River Compact, require extensive public works to store and rearrange the river's flow. In essence, the Colorado River and its tributaries flow into a system of dams, canals, ditches, dikes tunnels, pumps, turbines, and measuring devices connected by regulated streams—a huge plumbing system designed to manipulate water in a futile attempt to satisfy all who claim it. [125]

Colorado River Compact. A deal worked out between seven western states designed to stop the wrangling about who gets how much of the water that drains off the Colorado River watershed, including the San Juan Basin.

In the arid West, water is the key to development and wealth. In 1922 the states that use Colorado River water got together at Santa Fe and agreed on how much of the Colorado's water could be used by the Upper Colorado River Basin states—mainly Wyoming, Colorado, Utah, and New Mexico—and how much should flow to the Lower Basin states of Arizona, California, and Nevada.

The deal they celebrated was based on the assumption that the "normal" annual flow past Lee's Ferry, a river crossing in Arizona near the Utah border, was 16.4 million acre-feet. Codified by Congress in 1928, the agreement calls for the Upper Basin states to let 75 million acre-feet flow past that point every ten years. (Later estimates have put the annual flow at 15 million acre-feet, adding to the controversies.) The ten-year clause lets the upper states store water during high-flow years for release during low-flow years. The compact gave rise to the Colorado River Storage Project, which helps the upper states meet their obligation.

But the arrangement said nothing about how the Upper Basin and Lower Basin states would divide the water among themselves. The Upper Colorado River Compact of 1948, the Colorado River Storage Act of 1956, and the Colorado River Basin Project Act of 1968 followed to deal with that and other unresolved questions. [125]

C

Colorado River Storage Project (CRSP). A series of water-storage and related works in the Upper Colorado River Basin. The Upper Basin includes the west half of Utah, the southern tip of Nevada, some of Wyoming, the west slope of Colorado, a splotch in northwestern New Mexico, and a small section of northeastern Arizona. Among the project's four storage works is the Navajo Storage Unit, stretching across the Colorado– New Mexico line in the San Juan Basin. (The other three are the Glen Canyon, Flaming Gorge, and Wayne N. Aspinall units.)

The CRSP sprang from the Colorado River Compact, which stopped some of the fighting among the states over Colorado River water. In addition, the dams generate and sell electricity; part of the income goes to help pay for other CRSP projects. [125, 257, 265]

Colorado Territory. In April 1859 a bunch of gold miners got together at Richens "Uncle Dick" Wootton's Tavern on Cherry Creek—now part of Denver—and brazenly declared their own "State of Jefferson." Their act had no legality, but it laid the groundwork for a congressional argument about setting up a new territory. The politics of slavery, transcontinental railroad routes, and territorial self-government stalled the debate until the threat of southern rebellion forced Congress to look at the West with some urgency. It organized Colorado Territory in February 1861. [78, 117]

Colorado Trail. A nonmotorized recreation trail that stretches 469 miles from Durango to Denver. Volunteers, supported by the Forest Service, constructed this pathway starting in 1975. The trail starts by going up Junction Creek off La Plata 204 two miles north of Durango proper. It winds over Kennebec and Molas Passes before it leaves the basin by crossing the Continental Divide south of Stony Pass. On its way to Denver the route passes through seven national forests, six wilderness areas, five major river systems, and eight mountain ranges. The trail is maintained through an alliance of the Colorado Trail Foundation and the Forest Service. [70]

Concord Coach. The stagecoach that went from Animas City south to Farmington via Durango and La Posta twice a week in 1881 was likely a Concord Coach. It was proud to be the most elegant as well as the most efficient coach in the West.

Its chassis, wheels, and running equipment were always yellow, but its body could be scarlet or green. With a twelve-foot wheelbase carrying well over a ton, the basic model sold for about a thousand dollars. Plush linings coddled its passengers, and adjustable canvas or leather curtains adorned the unglazed windows. Leather springs cushioned the ride while candle lamps lit the twelve-passenger interior. You could take your place on any of its three upholstered benches, and if all the seats were taken, you could always ride on the roof. [223, 251]

Conestoga wagon. When the H. W. Cox and Alf Graves families coaxed their seven yoke of oxen and three-ton loads down the Animas Valley to Bondad in 1876, they were likely pulling a Conestoga-type wagon—the original "prairie schooner."

Wainwrights of Pennsylvania's Conestoga Valley started making these behemoths in the 1700s, but the market was mediocre until the California Forty-Niners used them to span the continent. This demand prompted other companies—primarily Studebaker and Lathans—to get into the act. Perhaps the vehicle got the name "schooner" from its concave wagonbed that swept up fore and aft like a real schooner's prow and stern. Then, too, some wagonboxes were watertight for fording or floating streams.

Evidently the Coxes' wasn't, for they prudently built a skiff to float their provisions across the Animas. Later they built a road and retrieved the wagons. [194, 223]

Continental Divide. The spine of the Rocky Mountains sends water from its west slope to the Pacific Ocean and, in the Southwest, off its eastern gradient to the Gulf of Mexico. Along most of its north-south traverse down the North American continent it follows a gentle course, but on the east edge of the San Juan Basin it travels a serpentine route like a meandering river in reverse relief.

Perhaps no other section of the divide is more dramatic than its division of watersheds in the rugged San Juans, a few miles southeast of Howardsville, where Stony Pass sends water west to the Colorado River and east to the Rio Grande. By the time these waters reach their respective seas (if not siphoned off to support civilization), they have embodied the essence of the Southwest. [236]

Coronado, Francisco Vásquez de (1510–1554). This Spanish explorer came north through New Mexico into present-day Kansas between 1540 and 1542 looking for the Seven Golden Cities of Cibola, Native American villages falsely rumored to be fabulously rich. Although he passed south of the San Juan Basin and got no closer than present-day Santa Fe, he laid the way for future Spanish colonization. [217]

Cortez, Montezuma County, Colorado. In the 1880s the Montezuma Valley Irrigation Company devised a scheme to irrigate the valley from the Dolores River, giving rise to an influx of construction workers, farmers, and merchants; the settlement of Cortez; and the city's incorporation in 1902. The town was laid out in 1886 on land owned by James W. Hanna, co-owner and manager of the irrigation company. In 1887 irrigation-related construction was in full swing, and hundreds of men were in town to work on the project. By 1900 the place had all the features of a settled community: post office, school, churches, courthouse, and newspaper.

The city has been a center of commerce for Montezuma County through its eras of fruit growing, dryland cultivation, uranium prospecting, oil production, dam building, tourism, and irrigated farming.

Cortez (1990 pop. 7,284; elev. 6,200 feet) is at the junction of U.S. 160 and U.S. 666, forty-four miles west of Durango. [50, 182]

cottonwood. Although it has little or no commercial value, this tree is good for holding stream banks and some inventive uses. Pioneers boiled its bark to make astringent tea. The basin is within the range of both the plains and Rio Grande cottonwoods, but the narrowleaf is most abundant. Since cottonwoods like low, moist ground, their habitat is enhanced by agrarian irrigation systems. [233]

Counselor, Sandoval County, New Mexico. Located three miles west of the Jicarilla Apache Reservation, this community grew up around the trading post built by Jim Counselor in the winter of 1931–1932.

Counselor is fifty-four miles southeast of Bloomfield on New Mexico 44. [87]

counties. Some call them state administrative units, others say they are institutions of local government. In either case, they are descendants of the English shire system, transplanted to Virginia. When they were arbitrarily set up as divisions of a territory, they normally spread over vast regions. As areas became more populated, territorial and state legislatures divided them into smaller units. For example, New Mexico's San Juan County was once part of Rio Arriba County. San Juan County, Colorado, once belonged to La Plata County. Officials who drew county boundaries normally ignored topographic features and cultural patterns. The way county boundary's crisscross the Continental Divide demonstrates their preference for survey lines.

Six counties lie wholly, or to a great extent, within the San Juan Basin as defined for this book. Excluding two small fractions of Rio Grande and Conejos Counties in Colorado, here are the six main counties of the basin, along with their county seats, 1990 populations, and areas:

Archuleta (Colorado)—Pagosa Springs; 5,345; 1,364 sq. mi. (50 sq. mi. outside the basin)

Dolores (Colorado)—Dove Creek; 1,504; 1,029 sq. mi.

La Plata (Colorado)—Durango; 32,284; 1,691 sq. mi.

Montezuma (Colorado)—Cortez; 18,672; 2,097 sq. mi.

San Juan (Colorado)—Silverton; 745; 392 sq. mi.

San Juan (New Mexico)—Farmington; 91,605; 5,942 sq. mi.

Four other counties have territories in the basin. Their county seats, estimated basin populations, and basin areas are as follows:

Hinsdale (Colorado)—Lake City; 0; 300 sq. mi.

Mineral (Colorado)—Creede; 0; 200 sq. mi.

Rio Arriba (New Mexico)—Tierra Amarilla; 3,000; 2,920 sq. mi.

Sandoval (New Mexico)—Bernalillo; 200; 450 sq. mi.

cowboys. We don't know how this term for cattle herders got started, but we do know the myth, if not the reality, of these riders of the range. Mountain men, homesteaders, prospectors, scouts, and military commanders emerged from the pioneer West, but none caught the public's imagination as did the cowboy. Hollywood gives this bigger-than-life hero a steady disposition with an obedience to the Code of the West, but he was often worse than just a rowdy.

In the basin he was likely to be an employee of the Two Cross Ranch, a large spread along the La Plata River at the Colorado–New Mexico line. Saloon brawler Tom Nance was one of its boys.

Good, bad, or indifferent, the cowboy is an icon not just for the West or for an era. He has proven to be an international emblem with amazing staying power. [85, 89]

coyote. To many Native Americans this canine is the smartest animal on earth. Perhaps no other wild animal evokes more admiration—and scorn. Coyotes range in color from near black to off-white; those in the basin are light-colored with still lighter throats and bellies. They stand about eighteen inches high at the shoulder and weigh up to fifty pounds. Ranging throughout North America, coyotes like open space best, but they thrive in mountains and adapt to cities as well. While normally elusive in the wild, in urban areas they can become aggressive enough to attack pets. They rarely bother people, but if you want one to go away, just pretend it's your neighbor's cur and yell and throw something at it. Coyotes have long been shot, poisoned, and trapped through government-subsidized eradication programs. But the coyote is still out there in most parts of the basin, yipping in the night. [138]

Crow Canyon Archaeological Center. A not-for-profit association for research and education located near Cortez. Using private and public lands with the cooperation of the Archaeological Conservancy and public agencies, the center conducts on-campus courses and workshops for amateurs, students, and professionals. Experienced archaeologists, many with advanced degrees, make up the staff. Started in 1984, the center provides a lodge and housing for participants.

You can reach Crow Canyon by going two miles north of Cortez on U.S. 666, turning left on Montezuma L, and left again on Montezuma 23. [141]

Crystal, San Juan County, New Mexico. The U.S. Army's 1849 Navajo Reconnaissance Expedition under Colonel James Macrae Washington found this place, with its red rock buttes interspersed with mountain greenery, to be an ideal camp site, except in winter when it is one of the coldest places on the present-day Navajo Reservation. It is not unusual to find a two-foot blanket of snow covering the area.

The Crystal Trading Post is unusual among Navajo communities due to its high elevation. Trader John B. Moore, when he bought the post in 1896, overcame its adverse climate by employing Navajo weavers to make rugs during months of winter isolation. He then took advantage of his location on the west slope of the Chuska Mountains' Washington Pass and established himself as a master trader of fine rugs.

Crystal (elev. 7,550 feet) is sixty-nine miles south of Shiprock by way of U.S. 666 and New Mexico 134. [71]

Cunningham Pass. Lying east of the creek of the same name, which flows from the south to join the Animas River, this was a route for prospectors entering the basin in 1861. Operators of the Highland Mary Mine improved it as a route to their holdings, and by 1871 wagon companies were using it for general freight transport. Because this pass is 500 feet lower than Stony Pass, less than two miles north, it was favored by many early shippers.

Cunningham Pass (elev. 12,180 feet) is accessible by hiking to the northwest edge of the Weminuche Wilderness. [219]

dance hall. Foot stamping provided a rare social outlet for pioneers, and a dance hall was a priority in frontier communities. Lacking such a facility, they were likely to use any structure available, sometimes a saloon. As the frontier matured, these dance halls became licensed and were called hurdy-gurdy houses. At Parrott City, near the head of La Plata Canyon, the townspeople used the courthouse. The dancers were so determined to enjoy themselves that they sometimes moved their shindig upstairs to avoid hooligan harassment. [166, 223]

deer. The mule deer in the basin-like habitat that sometimes includes residential areas. Their ropelike tail, evenly forked antlers, and large ears distinguish them from the white-tailed deer that live east of the Rockies. Since deer browse on woody vegetation and eat little grass, they do not compete seriously with livestock and wapiti (elk). They do, however, like the ornamental plants of the suburbs. Deer are sought by mountain lions, coyotes, and hunters but also fall victim to motor vehicles and packs of feral dogs. [135]

Dempsey, Jack (1895–1983). The boxer who knocked down his old friend, Andy Malloy, in a ten-round match at the Gem Theater, northwest corner of Tenth Street and Main Avenue in Durango, October 7, 1915. (A mural painted in the 1980s on the southwest corner falsely depicted the fight taking place on that quadrant.) "Kid Blackie," as Dempsey was called at the time, went on to become known as the "Manassa Mauler" and win the heavyweight championship.

Born at Manassa in south-central Colorado, William Harrison Dempsey started prizefighting while working in Utah and Nevada mining and timber camps. Four years after his Durango bout, he took the championship at Toledo, Ohio, in 1919 by fighting Jess Willard from his neighboring state of Kansas. After the fight Dempsey was greeted with the news that his manager had bet the $27,000 purse—that he was confident of winning—that Dempsey would beat Willard in the first round. Willard went down in the first but was saved by the bell. For taking the championship, the Manassa Mauler received not a cent. [116, 229]

De-Na-Zin Wilderness. According to legend, a large flock of cranes came to rest and feed here before continuing their migration, and that's why the Navajos gave it their term for "standing crane." Set aside by Congress in the San Juan Basin Protection Act of 1984 (with allowances for some mining and grazing), this unit of the National Wilderness Preservation System includes thirty-eight square miles of harsh, arid land that has resisted civilization.

Eons ago vegetation and reptilian creatures thrived in a semitropical environment here. Now the area lies in the high desert of the San Juan Basin. Weather and other forces slowly eroded its three geological formations—the Kirtland Shale, the Ojo Alamo Sandstone, and the Nacimiento Formation—to produce the shades of rust, gray, red, black, and white of its mesas and badlands. Petrified logs, dinosaur bones, and fossils evidence the region's past.

This wilderness has five distinct habitats: piñon-juniper, sandy washes of badlands, rolling grasslands, sandstone-capped mesas, and a remnant population of Ponderosa pine. Its badlands support little animal life, except snakes like the prairie rattler and lizards. But its piñon-juniper and grassland areas shelter animals common to those vegetation types. Hawks, prairie falcons, and golden eagles prey on that wildlife.

Managed by the Bureau of Land Management, the west edge of De-Na-Zin is twenty-two miles south of Bloomfield on New Mexico 44 and nine miles west from El Huerfano Trading Post on San Juan 7500. [150]

Denver and Rio Grande Railway (D&RG). The line that impelled the San Juan Basin's early growth. It was also the first, largest, and longest-lived major American railroad to use a narrow gauge.

Chartered by William Jackson Palmer in 1870 as the Rio Grande Railroad and Telegraph in New Mexico, it set out as a main line down the front range of the Rocky Mountains and along the Rio Grande to El Paso, then branched into the mining areas of the San Juan Basin. It entered the basin near Monero in 1881, passed through Lumberton and Dulce, crossed into Colorado along the Navajo River, then laid a grade via Pagosa Junction, Arboles, Tiffany, Ignacio, Oxford, Falfa, and Grandview to arrive at Durango in July the same year. With the extension of its line

The Denver and Rio Grande Railway came to the basin in 1880, and railroads were the area's principal mode of transportation until after World War II. This train was getting ready to steam from Durango to Alamosa. Courtesy La Plata County Historical Society.

to Silverton a year later, the D&RG had eighty miles of track in the basin, linking the area's commerce to the rest of the nation.

Unlike the transcontinental roads that received huge federal land grants, the D&RG got only a two-hundred-foot right-of-way and small acreages for depots. It did, however, extract concessions from towns along the way as a condition of favoring them with its service. And a community that refused to grant such concessions was likely to face the consequences. Animas City declined to give the D&RG what it wanted, so the railroad simply established its own townsite—which became Durango—and put the depot there. Animas City faded and eventually became part of Durango. The D&RG brought its constructed San Juan Basin mileage to 129 when it completed the 49-mile line from Durango to Farmington in 1905. In 1908 it expanded by purchasing the 20-mile-long Rio Grande, Pagosa and Northern. Locals dubbed the run to Farmington "The Red Apple Flyer." Unlike all the other basin lines, it was laid standard gauge.

The wider gauge proved a detriment; it required transfer of Farmington's outgoing freight to narrow-gauge cars at Durango for shipment to farther points. In 1923 the company got rid of that expense by changing the Red Apple Flyer to narrow gauge.

During and after World War II uranium and vanadium production helped keep the D&RG running, but in 1952 the demise of the Rio Grande Southern, which had provided the D&RG a way to go north out of the basin, left parts of the D&RG isolated. The Farmington oil boom of the early 1950s brought pipeline freight to the line, but after 1956 the railroad became a losing proposition.

In 1969 the Interstate Commerce Commission let the D&RG abandon all its lines in the basin except the Silverton branch. When the D&RG sold that part in 1981, the new owner called it the Durango and Silverton Narrow Gauge Railroad, operating today as a popular tourist attraction.

Incidentally, the D&RG Railway by any other name is still the D&RG. Palmer incorporated the Utah portion of the line as the D&RG Western Railroad. The D&RG reorganized out of receivership in 1886 to became the D&RG Railroad. When it was sold from another receivership in 1921, it became the D&RG Western Railroad. [6, 22, 63, 208]

Deseret. The Mormons used this name for the state of the Union they advocated in 1849. It was to include a stupendous territory consisting of perhaps a sixth of the area of the modern contiguous forty-eight states. The San Juan Basin lay within the proposed state on its eastern edge. The desired region included most of modern-day Utah, Nevada, and

Arizona; much of California, New Mexico, Colorado, and Wyoming; and pieces of Idaho and Oregon.

Most of that territory had come into the possession of the United States at the conclusion of the Mexican-American War in 1848. Like other annexed regions, it had no regional and local government. The Mormons tried to fill the void by adopting their own state constitution and sending a representative to Congress, seeking statehood.

In 1850 Congress set up Utah Territory to include all of present-day Utah, most of Nevada, the western third of Colorado, and part of Wyoming—a region little more than half the size of the Mormons' proposed state. That territory was reduced again by half when Congress split off regions for Colorado and Nevada Territories.

The Mormons named their settlement and would-be jurisdiction from the Book of Mormon. When Congress gave the territory a different name, "Deseret" became obsolete but its meaning, "land of the honeybee," hung on to serve as a nickname and symbol for both Utah Territory and later the state. [90, 204, 217]

Devil Creek State Wildlife Area. With San Juan National Forest lands on its west and north, this 561-acre tract's elevation ranges from 6,720 to 7,600 feet. Devil Creek's willows and cottonwoods interrupt the area's mountain browse, oak brush, and piñon. The Colorado Division of Wildlife bought the property for turkey production and now grows winter wheat to enhance its range for turkey, deer, and wapiti. Bears also use the area. Peregrine falcons hunt rabbits, Abert's squirrels, doves, blue grouse, and band-tailed pigeons. Beaver ponds enhance the land's wildlife habitat.

Devil Creek is sixteen miles west of Pagosa Springs on U.S. 160 and two miles north on Forest Service 627. [250]

dogie. An orphaned calf, as in the ballad "Whoopie Ti Yi Yo, Git Along Little Dogies."

> As I was a-walking one morning for pleasure,
> I spied a cow-puncher a-riding along;
> His hat was throwed back and his spurs were a-jinglin',
> As he approached me a-singin' this song:
> Whoopee ti yi yo, git along little dogies,
> It's your misfortune and none of my own;
> Whoopie ti yi yo, git along, little dogies,
> For you know Wyoming will be your new home. [83, 223]

Dolores, Montezuma County, Colorado. A town on the Dolores River, from which the community got its name, born in 1891–1893 when the Rio Grande Southern Railroad laid track up the valley to Rico. When the railroad's attorney and other investors bought a homestead and laid out the townsite that would become Dolores, settlers left the Big Bend community a couple miles downstream and moved to the new location. The settlement grew up as a cattle and agricultural center, matured with lumbering, then took advantage of its location on the edge of the national forest and the McPhee Reservoir to attract recreation seekers.

Dolores (1990 pop. 866; elev. 6,936 feet) is twelve miles north of Cortez on Colorado 145. [49, 50]

Dolores County, Colorado. This county on the northern edge of the San Juan Basin grew from an 1879 mining boom. When the news got out that prospectors had found lead carbonates rich in silver up the Dolores River valley, miners swarmed in from the surrounding mining districts. In 1944 the county seat's move from Rico, which was born in the mining boom near the east end of the county, to Dove Creek, located at the opposite end, reflected the waning of mining activity and the growth of agriculture. Irrigation from the Dolores Project has fostered the county's beans, barley, and other grain industries.

Dolores County was cut from Ouray County (originally part of La Plata County) in 1881. County seat: Dove Creek; population (1990): 1,504; area: 1,029 square miles. [40, 132, 208]

Dolores lumber railroads. To serve the timber camps, railway and timber companies built railroads around Dolores. The New Mexico Lumber Company (also called the Montezuma Lumber Company), the Dolores, Paradox and Grand Junction Railroad, and the Colorado and Southwestern Railroad Company were among the investors. During the peak cutting years of 1902–1948, the lumber operations used about sixty miles of track. [100, 175, 176]

Dolores Project. McPhee Reservoir, which can store 381,100 acre-feet of water with a surface area of 4,470 acres, is the central feature of this Bureau of Reclamation project. Built in the 1980s, it extends ten miles along the Dolores River and branches up several creeks. The project's water irrigates land around Cortez and Dove Creek and generates electricity. The reservoir has fifty miles of shoreline at normal high water and an average surface elevation of 6,894 feet. The bureau, the Dolores Water Conservancy District, the Forest Service, and the Ute Mountain Ute Tribe administer various parts of the project.

You can get to the lake just south of Dolores on Colorado 184. [258, 265]

Dolores River. This river has the most unusual flow of any in the basin—and for its size, perhaps in North America. It flows both south and north. Starting near Lizard Head Pass east of Rico, the stream flows southwest to Dolores, then turns north to meet the Colorado River in Utah. Of its 230 flowing miles, about 100 are in the basin.

The Spanish called it Rio de Nuestra Senora de las Dolores, meaning "River of Our Lady of Sorrows." Legends tell us that the Dominguez and Escalante party named it during their 1776 exploration because the Utes drowned captives in it. A less dismal version says Juan María de Rivera titled it during his 1765 expedition when two of his party drowned trying to cross the stream at flood stage.

Anglo settlers started farming with the river's waters when they organized the Montezuma Valley Irrigation Company in the 1880s. That practice has continued ever since, and the Dolores Project now interrupts the river's flow just downstream from Dolores to form McPhee Reservoir. [49, 50, 104]

Dolores River Recreation Area. An area administered by the Bureau of Land Management along the Dolores River northeast of Dove Creek. Its key features are Mountain Sheep Point and the Dolores Canyon Overlook. The overlook is on Montezuma 10 about ten miles from Dove Creek. [149]

Dominguez and Escalante Expedition. A 1776–1777 journey by two Franciscan fathers through the San Juan Basin looking for a way to Monterey.

While the Founding Fathers were damning the British crown for its governing excesses, Spain was claiming a domain stretching from the Rocky Mountains to the Pacific. To secure its holdings, Spain needed transportation routes, so Fray Francisco Atanasio Domínguez and Fray Silvestre Vélez de Escalante started from Santa Fe in July 1776 to find one. Fray Francisco Garcés had already shown the way from New Mexico to California by going south of the Grand Canyon, but Dominguez and Escalante were searching for a better route and had other motives as well. They wanted to explore the territory and convert the Utes to Christianity. For part of their way they followed the Old Spanish Trail, entering the basin near modern-day Monero and leaving the area north of modern Dove Creek. As they wandered they learned a lot about the West and did their best to spread their religion among the natives. But heavy snows in Utah forced them back, and they never made it to Monterey. [30, 33, 57, 76]

D

Dominguez and Escalante Memorial Highway. Colorado 184 northwest from Mancos to its intersection with U.S. 666.

Dominguez and Escalante Ruins. The remains of an ancestral Puebloan settlement named after two Spanish Franciscans who explored the basin in 1776–1777. The ruins are at the Anasazi Heritage Center on Colorado 184 three miles west of Dolores.

donkey. See **burro.**

Douglas fir. Although this evergreen will grow as low as 4,000 feet elevation, in the basin it prefers the wetter upper reaches where it thrives up to 11,000 feet. Compared with the giants of the Pacific Northwest, which may reach a height of 200 feet, the Rocky Mountain version is smaller and less straight-grained. It grows more slowly, tolerates drought better, and likes more sun. Its redder wood cannot be worked as easily, but it is stronger and more durable. [233]

Dove Creek, Dolores County, Colorado. This town grew up with the cattlemen and homesteaders who settled the Dolores River valley around the end of the nineteenth century. Sustained by its agricultural economy, it took the county seat away from Rico in 1944. Farming continues to play a major role in its economy; it is the center of an irrigated area where grain and pinto beans are grown. The community calls itself the "Pinto Bean Capital of the World."

Dove Creek (1990 pop. 643; elev. 6,843 feet) is thirty-seven miles north of Cortez on U.S. 666. [50]

dragoons. To call these soldiers "mounted infantrymen" may appear contradictory, but the term is correct. The American Dragoons were organized in 1833 within the U.S. Army to protect overland trade routes. In the West, where Native American attackers were mounted, the dragoons needed horses too. To fill the ranks, recruiters enlisted foreign mercenaries and asked few questions. Their elaborate uniforms sported gaudy badges, brass buttons, and yellow braids on sky-blue trousers. When they operated in the West before military posts were set up, however, their frontier dress was less formal; in the field they wore buckskin and other practical clothing. Calvary units replaced the dragoons in 1861. [223]

dry-gulch. When Henry "Billy the Kid" McCarty ambushed and murdered Sheriff William Brady of Lincoln County, New Mexico, near Ike Stockton's saloon in Lincoln, he "dry-gulched" the lawman. When

Stockton picked off Aaron Barker in a wash ten miles north of Farmington, his deed was more literal: he shot his adversary in a dry gulch. [85, 223]

dude. A term that got started in the 1880s to describe derby-hatted city residents from the East. Thus, a "dude ranch" accommodates and entertains tourists who are, to say the least, not ranchers. Old hands, sometimes derisively, use this term to describe a "tenderfoot," a term from the California gold rush. The word may have come from *dudenkop,* German for "lazy fellow." [223]

dugout. A covered excavation dug into the ground or carved out of a hillside. The pioneers used such shelters on the frontier for lack of conventional building materials. Dugouts were sometimes the only available shelter for Native Americans confined to camps or reservations, such as the Navajos imprisoned at Bosque Redondo. [82, 223]

Dulce, Rio Arriba County, New Mexico. Now headquarters for the Jicarilla Apache Tribe and located on their reservation, this community started when trader Emmet Wirt moved his post there from Lumberton in the early 1890s. The Jicarillas had little to trade and no money to spend, so Wirt often sold them goods on credit. His books were filled with unpaid accounts. When a debtor died or when an epidemic swept away a whole family, his ledger gave mute testimony: "Paid by God."

Dulce (1990 pop. 2,438; elev. 7,192) is seventy-one miles east of Bloomfield on U.S. 64. [198]

Dunton, Dolores County, Colorado. This mining camp was a Johnny-come-lately in the nineteenth-century rush for precious metals, and although the town got started in the 1880s, it didn't get a post office until 1892. It peaked at the turn of the century. After mining faded, its hot springs became a tourist attraction in 1917. It is now a German-owned private resort.

Dunton is thirty-five miles northeast of Dolores via Colorado 145 and Forest Service 535. [40]

Durango, La Plata County, Colorado. In 1879, after William Jackson Palmer decided to run the tracks of his Denver and Rio Grande Railway up the Animas Valley to the mining camps around Baker's Park, his negotiators visited Animas City's community leaders. They sought concessions in exchange for giving the town a depot, but the town's officials offered no favors.

San Juan and New York Smelter in Durango prospered in the late nineteenth century to mill both precious and base minerals from ores dug out of the San Juans. In the 1940s and 1950s the site housed vanadium and uranium mills. Courtesy La Plata County Historical Society.

The railroaders then formed a subsidiary, bought land a couple of miles south on the line's planned route, and laid out a townsite. A new town was born.

The influx of miscreants, speculators, merchants, and opportunists that swarmed to the railroad's temporary terminus looking for a piece of the financial action proved too much for the county government to handle. Members of the more stable element incorporated a city so they could hire a marshal and get the garbage, human and otherwise, off the streets. When the county voters took the seat of government away from Parrott City and gave it to Durango in 1881, they assured the upstart city of the dominance it has experienced ever since. Merchants, bankers, and professionals abandoned Animas City and set up shop in Durango. When a mill moved to the city from Silverton and the D&RG began hauling ore down the valley from the mines, Durango matured from a boom town into a stable, industrial community. It became the basin's livestock shipping point as cattlemen and sheepherders drove their herds to the Durango railhead. In 1890 Otto Mears's Rio Grande Southern Railroad went west from Durango, enhancing the town's position as a transportation hub. Although with less severity than other parts of the mining West, depression hit Durango in 1893 as the price of silver collapsed.

But the mines to the north were good for more than just silver and the town used coal mining, agriculture, and tourism to sustain it through the early twentieth century. During World War II its economy got a boost from an upsurge in coal mining. Then came a vanadium mill; after the war it processed uranium. As war-related industries faded, the energy boom of the 1950s began; the search for oil brought exploration companies to town. And while Durango did not see the fivefold increase that Farmington experienced during the decade, its population did jump 41 percent. Then between 1940 and 1960 it doubled from 5,000 to 10,000. In 1956 the town's economy received the stabilizing influence of Fort Lewis College as the school moved from its old army-post site south of Hesperus to a mesa above Durango.

The community's economy now relies on tourism (attracted partly by the D&RG's descendant, the Durango and Silverton Narrow Gauge Railroad) and recreation backed by education, light manufacturing, agriculture, general retailing, and a steady stream of newcomers attracted by the town's lifestyle.

Durango (1990 pop. 12,430; elev. 6,512 feet) is forty-nine miles north of Farmington via U.S. 550. [22, 110, 113, 208]

Durango and Silverton Narrow Gauge Railroad (D&SNG). In 1981, when Floridian Charles E. Bradshaw Jr. bought the Denver and Rio Grande Western Railroad's line, running from Durango to Silverton,

that's what he named it. In 1969 the D&RG had abandoned its lines going south and east from Durango, but the Interstate Commerce Commission made it keep running the train to Silverton. As a result, the
railroad now had a tourist attraction cut off from the rest of its system.
This isolation and the company's lack of enthusiasm for the tourism
business gave reason to sell the line to Bradshaw. The D&SNG carries
some 200,000 tourists annually on its forty-six-mile run from Durango
to Silverton through the spectacular Animas River canyon. [63]

Durango Fish Hatchery. This operation goes back over a century; its
first building went up in 1893. The 1.5 million trout and kokanee
salmon fingerlings and 200,000 catchable-size trout raised here yearly
eat over a hundred tons of food. Wildlife biologists stock more than 120
Colorado streams and lakes with the fish. Operated by the Colorado Division of Wildlife, the hatchery features a visitor center and wildlife museum. It is on the east side of the 1600 block of Main Avenue, Durango.
[136]

eagle. It is not unusual to see the national emblem of the United States, the bald eagle, scouting the Animas River through Durango or the Pine River near Ignacio. Eagles use the same nest but add materials every year. Some nests become five feet high. Two species of this bird of prey live in the West, the golden eagle and the bald eagle. Larger than a hawk and without the bald head of a vulture (bald eagles aren't bald), both species can have wingspans over seven feet and hunt during the daytime. They are legally protected. [217]

Echo Canyon Reservoir State Wildlife Area. Situated in a mountain setting of ponderosa pine and native grasses and at an elevation of 7,100 feet, this 118-acre reservoir was built by the Colorado Division of Wildlife in 1968 for public fishing. The lake features largemouth bass, channel catfish, green sunfish, yellow perch, and stocked rainbow trout. Some waterfowl frequent the area.

Echo Canyon is four miles south of Pagosa Springs on U.S. 84. [250]

Edith, Archuleta County, Colorado. Lumberman S. M. Biggs started this timber camp on the Navajo River just north of the Colorado–New Mexico line in the 1890s. At its peak the mill was reported to be the most complete in the area and had a capacity of 60,000 board-feet per day. It was powered by electricity even before nearby Pagosa Springs enjoyed that luxury. From Edith the Rio Grande and Pagosa Springs Railroad hauled lumber south through Lumberton, then east over the Continental Divide to Chama.

In 1900 the town was at the heart of a legal battle over the state line. Losers in the Archuleta County election tried to have the Edith votes thrown out, contending that its residents lived in New Mexico. The judge refused to hear the case, but the boundary question remained until the 1950s.

Edith is south of Pagosa Springs eighteen miles on U.S. 84 and six miles on Archuleta 359 (locals call it Coyote Park Road). [91]

El Camino Real. Named "the royal road" by the Spanish, no route did more to help them colonize New Mexico during the eighteenth century. After splitting from Old Spain in 1821, New Spain (Mexico) used it as a principal military and trade route from Chihuahua up the Rio Grande to Santa Fe. At the road's terminus, Santa Fe became the center of commerce as well as the seat for government and military command. The Santa Fe Trail effectively extended the trail to Missouri while the Old Spanish Trail came from Santa Fe north through the basin on its way to California. [28, 122]

Electra Lake. A reservoir between Durango and Silverton built in 1906 to generate electric power. The project proved to be too ambitious; in 1915 it still had a market for only 40 percent of its capacity. Its owner, the Animas Power and Water Company, went into receivership. The lake receives water from Cascade Creek through a flume, then drops the water to a generating plant at Tacoma on the Animas River. The reservoir is open to public fishing.

The road to the lake goes east off U.S. 550 twenty-three miles north of Durango. [5, 95]

Elitch, Mary (1856–1936). Her husband, John Elitch, tried to be a restauranteur in the basin before he founded Elitch Gardens in Denver. In 1880 John opened a restaurant for fine dining on Second Avenue in Durango. A good idea in San Francisco, where he had run a restaurant for the theater crowd, it was too expensive for Durango. So he opened a more run-of-the-mill "chop house" on a side street. It was apparently not a financial success either, because when the Elitchs moved to the Denver scene, they were broke.

There John worked in restaurants until he had the capital to open his own, the Tortoni, on Arapahoe Street. To raise vegetables for his eatery, he bought acreage in northwest Denver. When P. T. Barnum gave him surplus circus animals in 1890, Elitch's Zoological Gardens was born. John died in 1891, but Mary carried on to develop a premier entertainment center. Over the years the gardens have been home to a theater of national repute, a ballroom for the big band era, and a baseball team. Mary let young Douglas Fairbanks Sr. scrub the stage for a ticket and lured such big names as Sarah Bernhardt to perform.

Moved and rebuilt in 1994, Elitch Gardens is now one of the West's premier amusement parks—and it all happened because the Elitches didn't make a go of it in Durango. [47, 66, 193, 195]

elk. See **wapiti.**

Engelmann spruce. Almost certainly the most dramatic tree in the Rockies, this conifer grows at elevations between 8,000 and 12,000 thousand feet with spires up to 100 feet tall. How appropriate that it be named for George Engelmann, the nineteenth-century German-born conifer authority. Although good for general building purposes, its wood is best for uses that need its straight stem—utility poles. Left free of avalanches, severe fire, or clear cutting, a spruce forest is likely to renew itself indefinitely, for its seeds take hold even in deep shade under dense growth. Then, as decay, windfalls, or chain saws take the older trees, seedlings spurt upwards in the newfound sun. [233]

Engineer Pass. Otto Mears built a road through the basin in 1877 that served as the Lake City–Animas Forks stage route. From where the stage changed horses you can see Engineer Mountain to the south. Now part of the Alpine Loop Byway, the road goes through American Flats, known for its profusion of wildflowers. The pass separates the watersheds of Bear Creek and Henson Creek.

Engineer Pass (elev. 12,800 feet) is a couple of miles north of Animas Forks. [219]

Escalante. See **Domínguez and Escalante Expedition.**

Eskridge, Dyson (ca. 1860-?). Outlaw co-leader with Ike Stockton of what became known around Durango and Farmington as the Stockton-Eskridge gang. His brother, Harg, also ran with the gang.

When the two siblings were running cattle in the Farmington-Aztec area of New Mexico in 1880, Dyson and a couple of other rowdies harassed a Christmas Eve party at Francis M. Hamlet's cabin four miles up the San Juan River from Farmington. In a shooting episode, one of the three killed a reveler. This put Dyson on the wrong side of a bunch of vigilantes known as the Farmington Stockmen's Protective Association. Dyson and Harg took refuge in Colorado. There the brothers joined up with Ike Stockton, who also had a grievance against the Farmington vigilantes—they had killed his brother, Port. This alliance gave rise to the Stockton-Eskridge gang.

Members of the outlaw bunch forayed into New Mexico to retrieve Eskridge's left-behind cattle and get revenge for Port's murder. They dry-gulched one of the killers, rustled cattle indiscriminately, and spread general anxiety among the populace, especially south of the Colorado–New Mexico line.

Dyson and Harg's older brother, Dow, ranched in the San Luis Valley. Sometime during his kid brothers' hell-raising, he rode over to the basin intent on straightening them out. He apparently failed.

The night of August 24, 1881, Dyson was with Bert Wilkinson and two other sidekicks at Silverton when Marshal David Clayton "Clate" Ogsbury approached the bunch. Wilkinson shot and killed him. Harg and Wilkinson fled the town and hid out together. To collect a reward, Stockton went after Wilkinson and found him camped out with Eskridge. On a ruse by Stockton, Eskridge left the camp. He was saved having to take sides when Stockton betrayed Wilkinson.

Following Stockton's death a few weeks after the double-cross, Dyson and Harg left the San Juan country. [1, 3, 85, 202]

Eureka, San Juan County, Colorado. This mining camp got started around 1872, but there is no reason to believe it deserved its name (Greek for "I have found it") more than any other settlement in Baker's Park. A townsite company made the place official in 1874, and the town prospered as long as the mines produced.

What's left of Eureka is seven miles east of Silverton on Colorado 110 and San Juan 2. [96, 115]

Falfa, La Plata County, Colorado. This station on the Denver and Rio Grande Railway served settlers on Florida Mesa southeast of Durango. It was the scene of a tragic accident in the 1890s. As they are prone to do, a team of horses became frightened by the train and bolted. Mrs. George Davis was run through by the wagon tongue.

Falfa was on present-day La Plata 221 a little east of Colorado 172. [170]

Falls Creek Archaeological Area. Basketmakers lived here sporadically from 300 B.C. until 700 A.D., and paleo-natives may have occupied the site 9,000 years ago. Rock shelters built by these ancestral Puebloans helped preserve materials found nowhere else in the Southwest. It was discovered in 1934 and acquired by the Forest Service in 1992 with the help of the San Juan Basin Archaeological Society and other organizations.

The site is four miles northwest of Durango via La Plata 204 and La Plata 205. [157]

fanning. The act of cocking and releasing the hammer of a single-action revolver by brushing it with the palm of the hand. Such a firing method has little accuracy and occurred in the West about as often as a quick-draw duel in the middle of town—not more than two fanning instances are on record (except in movies). [223]

Farmington, San Juan County, New Mexico. The largest city in the San Juan Basin, situated where the La Plata and Animas Rivers flow into the San Juan; hence, in their language, the Navajos call it "three rivers" or "between rivers," depending on which Navajo you ask.

Europeans who started the community in 1876 found a dry land of mesquite and sagebrush but readily irrigated the rich earth in the lowlands to grow fruits and vegetables. By 1892 their produce was winning awards at the Albuquerque Territorial Fair and they were exporting fruit by the hundreds of tons.

In 1885 the town was big enough to attract the vendor of a stereopticon lantern show. The effete easterner used the one-room schoolhouse as a theater. As he projected religious pictures, a group of cowboys at the rear of the room emptied their six guns at a character portrayed on the canvas. In a scene fit for Hollywood, the showman jumped through a window, taking the sash with him.

The townsite was laid out by Judge Samuel Webster to compete with Junction City, which lay just across the Animas River; community leaders incorporated the place as a town in 1901. The Denver and Rio Grande Railroad came south from Durango in 1905 to enhance the fruit export business. When the Works Project Administration (WPA) improved the road to Albuquerque in the 1930s, Farmington became even more accessible and until the 1950s lived up to its name as a farming community. Then the development of energy resources—coal, oil, and natural gas—swelled its population fivefold in a decade. In 1950 the building department issued $600,000 in building permits; in 1952 that number ballooned to almost $1,400,000. To build streets, shopping centers, industrial facilities, and housing for the 20,000 people who swarmed in during the boom, bulldozers uprooted the orchards, scraped off prime bottomland, and forever changed Farmington's appearance. What had been a quiet farming community quickly became a humming energy center.

That energy-related commerce gave a shot in the arm to the D&RG and kept the Farmington line running until 1956. Since then, energy demands, federal policies, and international oil supplies have pushed and pulled Farmington's economy, but it has remained the principal trade center for the San Juan Basin and the Four Corners area.

Farmington (1990 pop. 33,997; elev. 5,395 feet) is forty-nine miles south of Durango via U.S. 550 and 183 miles northwest of Albuquerque. [3, 15, 51, 85, 190, 192, 204, 215, 247]

fast draw. Some professional coaches of Western movie stars have been timed drawing a revolver in one-fifth of a second, but few real gunslingers bothered to develop such a talent. It was safer to dry-gulch an enemy or shoot from behind a barn. [223]

fence wars. The cattleman took advantage of three business breaks to become a tycoon: cheap cattle, low operating costs, and free land. Until the late 1880s, cattle roamed the public lands unrestricted. Settlers threatened to take away the last of the three advantages.

Cattlemen responded by stringing barbed wire, but they also had other reasons to build fences: to secure "their" range and water holes; protect purebred stock from impregnation by inferior bulls; and exclude nomadic sheep flocks and cattle herds. (Basin cattlemen protested bitterly about the intrusion of itinerant herds passing through from Texas and Arizona.) An epidemic of fence cutting followed, both brazen and under cover of night. Hundreds of fence-cutting reports inundated the General Land Office. Violence soon followed.

The laws of unintended consequences took over when, under an 1885 congressional act, President Grover Cleveland ordered all fences on public lands removed. As the fences came down, the General Land Office failed to replace these unlawful limits to public land use with any legal checks. Open range wars replaced fence wars. Cattlemen-sheepherders clashes got worse. Severe overgrazing followed.

In the late nineteenth century, a conservation movement convinced Congress to enact laws that curtailed some of the most blatant abuse of the public domain, including illegal grazing. But all the federal agencies combined could not effectively police the quarter million square miles of public lands. Illegal fencing and grazing were widespread until the Taylor Grazing Act of 1934 organized the public grasslands into grazing districts. The act has been criticized for putting the public range under the control of stockmen's committees. But it did, at last, set a workable framework for public range management. That system continues today. [54, 105]

firewater. Native Americans' term for strong drink, primarily whiskey. The name comes from the fact that high-proof liquor will burn when tossed on a fire. [223]

Fish Creek State Wildlife Area. This 309-acre tract along a mile of Fish Creek lies among riparian meadows of alders, cottonwoods, and willows framed by side hills of aspen, spruce, and fir. Its elevation is between 8,900 and 9,200 feet. In addition to its trout fishery, the area provides habitat for deer, elk, and bear. It is surrounded by San Juan National Forest lands.

Fish Creek is north of Dolores seven miles on Colorado 145 and seventeen miles via Forest Service 535 and an access road. [250]

Flora Vista, San Juan County, New Mexico. Pioneers settled here in the 1870s, stopping on the southeast bank of the Animas River. When the river changed course and threatened the community, they moved to the northwest bank. That's where the community remains.

Flora Vista (1990 pop. 1,021) is five miles southwest of Aztec on U.S. 550. [3]

Florida, La Plata County, Colorado. This frontier stage station served lines that linked Animas City with Pagosa Springs to the east and Tierra Amarilla to the south. After the Denver and Rio Grande Railway came through in 1881, the place served as both a stage and railroad station.

Florida was on the Florida River a mile south of what is now U.S. 160. [248]

Florida Project. Lemon Dam is the principal feature of this public work, built by the Bureau of Reclamation on the Florida River in the early 1960s. Its reservoir holds 40,146 acre-feet of water under a surface area of 622 acres. Half a mile wide and three miles long, the reservoir stores irrigation water for Florida Mesa, the tableland between the Animas and the Pine Rivers southeast of Durango. The Florida Water Conservancy District operates the dam, and the Forest Service provides recreation grounds around the lake.

You get there by going fourteen miles east out of Durango on La Plata 240 (locally known as Florida Road) and north on La Plata 243. [259]

Florida River. This stream starts in the Weminuche Wilderness, where it feeds Durango's municipal water reservoir before flowing into Lemon Reservoir. Its name is Spanish for "flowering." The river flows into the Animas River near Bondad.

fort. Military posts in the frontier West were commonly called "forts." The term includes arsenals, barracks, batteries, camps, cantonments, and depots. Of the more than 175 set up, primarily to subdue Native Americans, three are of special interest due to their location in or near the San Juan Basin: Fort Lewis, which had two locations in the basin; Fort Defiance in Arizona; and Fort Lowell in New Mexico. [216]

Fort Defiance. Set up in 1851 about thirty-five miles northwest of present-day Gallup, New Mexico, to control the Navajos, this was the first U.S. military post in what became Arizona. The garrison moved from the fort in 1861, and the facility became the Navajo Indian Agency in 1868. [216]

Fort Lewis. The "Meeker Massacre" at the White River Agency 200 miles north of the San Juan Basin and the threat of Native American uprisings prompted the army to move this post to its final location on the La Plata River. One of over 175 forts in the West, it was first set up as a camp at its original Pagosa Springs site in October 1878. It became a fort in July 1880.

The 1879 White River incident sent a shudder through the Anglo settlers of Ute territory. General Philip H. Sheridan, commander of the army's Division of the Missouri, responded to the perceived threat in the basin by sending in the cavalry (and the infantry). In the fall of that year, 600 troops hoofed to Animas City (now part of Durango). They included the Ninth Cavalry, reinforced by troops from New Mexico and Indian Territory (now Oklahoma). Their mission was to allay fears that the Utes might attack the settlement. When the Utes remained peaceful, Sheridan moved the garrison to its permanent site ten miles farther west on the La Plata River. During construction at its new location, the post was called "Cantonment on La Plata."

From this post the troops affirmed Anglo-American dominance of Utes, Navajos, and Apaches throughout parts of Colorado, New Mexico, Utah, and Arizona. After the fort closed in 1891 and became obsolete for other government purposes, the site evolved into a setting for Fort Lewis College.

Fort Lewis was located where you now see the San Juan Basin Research Center, four miles south of Hesperus on Colorado 140. [114, 204, 208, 216]

Fort Lewis College. A school that graduated from an army post to a liberal arts college.

When troops were no longer needed to protect Anglo-Americans from attacks by displaced Native Americans in the Four Corners region, the army abandoned its fort on the La Plata River four miles south of Hesperus and turned the land and buildings over to the Department of the Interior, which used them for an Indian boarding school. The school was like many others set up out west (and some back east) to train young Native Americans to be like Anglos. Although some Anglos and Hispanics went to the school for convenience, most of the students were Apaches, Utes, and Navajos. Declining enrollment, rising costs, and vacillating government policies forced the school's closure. In 1911 the department transferred the facility to the State of Colorado for educational purposes with an agreement that American Indian students could attend the state institution tuition-free.

Under state ownership, Fort Lewis started as an agricultural and vocational high school but soon began offering junior college courses. In 1933, after affiliating with Colorado State College of Agriculture and Mechanical Arts, it became Fort Lewis A&M. The institution became independent under the State Board of Agriculture in 1948. By the time it moved to its present setting on a mesa overlooking Durango in 1956, the school had become a four-year liberal arts college. Colorado's "Campus in the Sky," as Fort Lewis calls itself, is now a fully accredited liberal arts college with over 4,000 students. It offers twenty-four degree programs in its schools of arts and sciences, education, and business administration. Political correctness slighted the school's background in the 1990s when "Skyhawks" replaced "Raiders" as its emblem, but the term "Fort" remains to reflect its military heritage. [114, 143, 208]

Fort Lowell. The army used this military post a few miles east of the basin on the Chama River from 1866 to 1869 to protect white settlers from Ute depredations. It was just south of Tierra Amarilla. [216]

Four Corners. Situated on the west edge of the basin, forty miles southwest of Cortez, this is the only spot in the nation common to four states. What process gave birth to this unique feature?

When the United States took over the part of Mexico ceded by that country at the conclusion of the Mexican-American War, it inherited a system of administrative units with east-west boundaries running all the way to California. Congress chose the north border of one of those units (called Taos), running along the 37th Parallel, as the dividing line when it split the region to form New Mexico and Utah Territories. When Colorado Territory was set up in 1861, the New Mexico–Utah line (37th Parallel) east of what is now Four Corners became the New Mexico–Colorado line. For reasons obscure, Congress set Colorado's western border as a line north from the Four Corners spot. Wary of eastern New Mexico's Rebel sympathies during the Civil War, Congress split off the Union-secure west half of the territory as Arizona and completed the cross by drawing a line south along the same meridian it had used for Colorado.

The term "Four Corners" also applies to an economic region around the survey mark. To the Four Corners Tourism Council, it extends about 160 miles in all directions while to businesses "serving the Four Corners" it may be a smaller area. [204, 247]

Four Corners Generating Station. Operated by the Arizona Public Service Company for a consortium of electric companies, this plant's 2,040 megawatts are transmitted to Phoenix, El Paso, and communities between. Trucks and rail cars carry coal to the plant from the nearby Navajo Mine. Crushers then pulverize the coal into powder. The plant mixes the powder with preheated air and injects it into boilers where it

Four Corners Generating Station near Waterflow burns Navajo coal to generate electricity for Phoenix and other cities. Its Morgan Lake serves the plant and also provides habitat for dozens of bird species.

bursts into flame and makes steam to generate electricity. Water for steam, and to cool the plant, comes from adjacent Morgan Lake.

When the plant started up in the 1960s, it was the largest coal-fired plant in the world. It originally had few pollution controls and daily spewed 450 tons of smog-inducing gases and 250 tons of soot into the sky above the Navajo Reservation. So large and intense was the plume that the first astronauts saw it from space. The plant now scrubs and filters its exhaust to meet state and federal standards, but it can still cast a haze during certain atmospheric conditions.

The plant is eight miles southwest of Kirtland via San Juan 6100 and Navajo 5086. [82, 134]

Four Corners Monument Navajo Tribal Park. One of eight operated by the Navajo Tribe, this park features a marker indicating the only place where four state boundaries converge. The slab displays the seals of Arizona, Colorado, New Mexico, and Utah.

It is forty miles southwest of Cortez on U.S. 160. [164]

Fremont, John Charles (1813–1890). An army officer, politician, and famous explorer who may have entered the San Juan Basin. He at least came close during his fourth expedition.

As a second lieutenant in the Army Corps of Topographical Engineers, he helped reconnoiter the Minnesota country before eloping with Missouri senator Thomas Hart Benton's daughter, Jessie. During his first three expeditions Fremont explored the Oregon Trail, the Great Basin, California, and other regions of the West, compiling maps as he went. His daring January 1844 winter crossing of the Sierra Nevada perhaps foretold a fourth, far less successful trek just over the Continental Divide to the east of the San Juan Basin.

Fremont helped the United States take California away from Mexico and became its first civil governor. In that position he was court-marshaled and found guilty of challenging the authority of General Stephen Watts Kearny. Although President James Polk remitted his discharge from the army, Fremont resigned anyway.

His fourth excursion, 1848–1849, brought him to the San Juans and the face of disaster. Backed by Benton and St. Louis businessmen, he undertook to find a railroad route from Bent's Fort, on the Arkansas River in Colorado, to California. This entailed crossing the Continental Divide somewhere above the headwaters of the Rio Grande, perhaps at Cochetopa Pass. Such a crossing on foot with pack mules is difficult, even in good summer weather. But in the face of doubts and warnings by resident Anglo mountain men and Utes alike, Fremont's thirty-member party started for the pass in the middle of December. And the winter was severe.

Fighting howling blizzards, snow drifts that could be a hundred feet deep in the canyons, and temperatures plunging to twenty degrees below zero, the company was reduced to eating mule meat. Fremont, his judgment impaired by altitude sickness, delayed a decision too long. (In his delirium he may have ascended Stony Pass and briefly entered the San Juan Basin.) When he finally did abandon the effort, it took the group twenty days to ascend Embargo Creek twenty miles and find the Rio Grande.

The party then divided. One bunch started for help. Disasters compounded as Fremont headed for Taos. Before the debacle was over, a third of the group was dead.

Fremont went on to become a California senator, Republican presidential candidate, and Arizona governor. He died in New York City forty-one years after his San Juan Mountains disaster. [45, 56, 172, 204, 221]

Fruitland, San Juan County, New Mexico. Over 1,000 Navajos under Chief Kastianna massed here in 1894 ready to defend a tribal member,

Nesh-Kai-Hay, who had pawned beads with trader Henry Hill. The two had quarreled over a redemption value. In an ensuing encounter Nesh-Kai-Hay shot and killed one of Hill's employees.

To summon help, Fruitland resident Clay Brimhall rode off in the dead of night for Durango, where he telegraphed Fort Wingate. Army troops arrived in time to block the Navajos and persuade them to surrender. Nesh-Kai-Hay was convicted of murder and imprisoned at Santa Fe.

Before the pioneers developed irrigation and a fruit-growing postmaster renamed the place, it was called Burnham, after Luther Burnham, the Mormon who went there from St. Johns, Arizona, to convert Navajos and preside over a new Mormon ward. By 1881 a number of Mormon families had settled to farm and mine coal for sale in Farmington.

Fruitland (1990 pop. 700) is a little south of U.S. 64 nine miles west of Farmington. [3, 85]

Gambel oak. This deciduous tree's brown-red fall colors on the foot-hills lend contrast to the greens of the conifers and yellow of the aspen, but as a prolific shrub it is a nuisance on the range. Chemicals only slow it down, and if you chop it out, the sprouts come back thicker than the original stand. Infested acreage is hard to reclaim.

The most common species of oak brush in the southern Rocky Mountains, this shrub, or small tree, can get fifty feet tall, but its height, like the shape of its leaves, varies widely. It often mingles with ponderosa pine at elevations of 5,000 to 8,000 feet, the upper reaches of its habitat.

Also called Rocky Mountain white oak, the plant is named for ornithologist William Gambel, who identified the plant in 1844 while traveling in the West with fur trappers. [235]

Gato. See **Pagosa Junction.**

Gavilan, Rio Arriba County, New Mexico. Located three miles west of the Continental Divide and the Carson National Forest boundary, this place has a name that means "sparrow hawk" in Spanish.

Gavilan is six miles north of Lindrith on New Mexico 595.

Gem Village, La Plata County, Colorado. A Bayfield gem and mineral merchant set up a colony for artisans here in the 1940s, two miles west of Bayfield on U.S. 160. The colony has gone, but the name remains.

general store. The frontier emporium that purveyed everything from butchered animals to axle grease. Many pioneers used this expression derisively because, given the transportation problems, they were "generally out" of goods. [3, 168]

Hubbard's Market, ca. 1904. Like other general stores, this Farmington emporium tried to stock everything from axle grease to yarn to sides of beef. *Left to right:* C. M. Hubbard, Bert Hubbard, and Harley Kite. Courtesy Farmington Museum, John B. Arrington Collection, 1980.14.74.

Geronimo (ca. 1825–1909). So far as we know, this most famous of native warrior-leaders never came to the San Juan Basin, but historians mark his final surrender in 1886 as the end of Native American hostilities. It took 5,000 soldiers ordered into Mexico by General Nelson A. Miles to finally convince this Chiricahua Apache and a couple dozen other holdouts to give up. He died a prisoner of war. [246]

Gladstone, San Juan County, Colorado. A mining settlement founded by the San Juan Reducing Company in 1878, this community had its post office closed three times before it finally declined beyond revival. As the processing center for the Gold King and other mines, it became the destination of the Silverton, Gladstone and Northerly Railroad in 1899. After the Gold King shut down and the railroad pulled up its rails in 1924, the place was without a purpose.

Gladstone is seven miles up Cement Creek via Colorado 110 north of Silverton. [63, 96, 97, 109]

Gobernador, Rio Arriba County, New Mexico. Other than the meaning of the name—Spanish for "governor"—the history of this community remains obscure.

Gobernador is thirty-eight miles east of Bloomfield on U.S. 64.

Goodnight, Charles (1836–1929). The cattleman who first brought Texas longhorn cattle to New Mexico and Colorado. Some of his animals were likely the earliest of this breed in the San Juan Basin.

After migrating to Texas from Illinois at the age of ten with his mother and stepfather, Goodnight became a ranger and scout. He got started in the cattle business with his stepbrother, John Wesley "Wes" Sheek (the father of James Sheek, who moved to Mancos), but took off on his own to herd cattle north. Unlike the drovers who headed for the Kansas railheads, Goodnight saw opportunities at the forts, Indian agencies, and mining camps farther west. With Jim Loving he trailed cattle up the Pecos River to Fort Sumner, New Mexico. The Goodnight-Loving Trail went past Denver on its way to Cheyenne, Wyoming, where it became a leg for the Montana Trail to Canada.

Using money from driving thousands of cattle north, Goodnight joined with Britisher John George Adair to form the JA Ranch. Together they ran 100,000 cattle on a million acres of the Texas Panhandle. Using Hereford bulls, they bred one of Texas's top quality herds and developed "cattalo" by breeding Angus with buffalo.

In his last years he lived on a small ranch and invested in Mexican mines. After a life of consuming tons of red meat and miles of cigars,

Charles Goodnight succumbed to a series of heart attacks at age ninety-three. [59, 221]

Graves, Alfred F. (ca. 1850–?). Although a Rio Arriba County deputy sheriff, this vigilante rode with the Farmington Stockmen's Protective Association when its gunmen went after Port Stockton and the killers of George Brown. Brown was gunned down at the Hamlet cabin incident on Christmas Eve, 1880. After the vigilantes did away with Stockton, some members of the group, in an attempt to lend an air of legitimacy to the incident, reported that Deputy Graves had fired the fatal shot.

Alf Graves was among the first white settlers in the Animas Valley. As late as the 1970s a sandstone house stood in the community of Cedar Hill along the west side of the Animas River five miles south of the Colorado–New Mexico line. On a slab under a diamond-shaped window was the figure "77"—the year Alf Graves had first arrived there with his in-laws, the Coxes. Together they had trailed up from central Texas 4,500 cattle and 100 horses. Hauling three tons of provisions, they drove the first wagon, pulled by seven oxen, down the Animas Valley. Graves made a claim on the north side of the Animas River near the modern highway bridge.

The next summer Graves returned to Texas and came back with his family, including twins Mary and Tilda and baby John. Their provisions, packed in covered wagons, included food for a year, farm implements, chickens, and a pair of hogs. Coming down the river at Twin Crossings, they faced a stream too high to ford. The answer was to get some lumber from Animas City and build a skiff, which they did. The boat handled their provisions but could not accommodate the Conestoga-style wagons. To retrieve the wagons, the Graves and the Coxes built the first crude road above the river at Twin Crossings. Knowing firsthand the difficulties of crossing the Animas, Graves later built a toll bridge near his place. We don't know if that was before he traded his claim for a Winchester.

Alf Graves remains an enigma. He was, by most accounts, an industrious pioneer and a responsible family man. Yet he could be irresponsible. Not only did he help the vigilantes dispose of Port Stockton, but he also took part in the basin's most reckless gunfight (at Durango on April 11, 1881) and joined a bunch of hooligans to harass Reverend Griffin when the minister came up from Texas to preach at Bloomfield.

Perhaps his later life was more consistent. He became Preacher Griffin's first convert. [85, 179, 194]

Grey, Zane (1875–1939). Western novelist who penned his most popular novel, *Riders of the Purple Sage* (New York: Harpers, 1912), while holed up in a log cabin at Dove Creek.

Born Pearl Zane Gray at Zanesville, Ohio, a city named after a maternal ancestor, he chose to alter the spelling of his last name and drop the first. His skill as a baseball player helped him gain acceptance into the University of Pennsylvania, where he became a dentist. While in dentistry he began writing, helped by his wife, Lina Elice, who had studied English at Hunter College.

His ambition to write about western subjects got a boost when he went west to help C. J. "Buffalo" Jones with a scheme to cross buffalo and cattle. From that experience he wrote memoirs that became *The Last of the Plainsmen* (New York: Outing, 1908). His Westerns, starting with *The Heritage of the Desert* (New York: Harpers, 1910), were simplistic stories of good versus bad. Many of his stories celebrated the coming of the railroad, the annihilation of the buffalo, and the removal of the Native Americans as a vital step for America's westward expansion.

In addition to novels, he wrote animal stories and the text for a comic strip, *King of the Royal Mounted.* His books sold over fifty million copies, became the plots for 130 movies, and served as the basis for *Dick Powell's Zane Grey Theatre* (CBS, 1956–1962). [47, 50, 242]

groundhog. See **marmot.**

Groundhog Reservoir State Wildlife Area. Located in Dolores County in open, high mountain ranchland and surrounded by forests of aspen, spruce, and fir, this lake sits at an elevation of 8,900 feet. When full, the reservoir has a surface of 668 acres with water depths up to ninety-seven feet. Although its waters are impounded for irrigation, a minimum conservation pool maintains the lake's fisheries. It features native cutthroat and rainbow trout.

Groundhog Reservoir is north of Dolores thirty-eight miles on Forest Service 526 and five miles on Dolores H. [250]

gunfight. The normal western gunfight did not involve the mythical fast draw in the center of town, of which fewer than a handful are recorded. Nor was it likely to be the prolonged shootout depicted by Hollywood. Nevertheless, the San Juan Basin produced a shooting spree worthy of any pulp Western or grade B movie. It happened on a Monday morning, April 12, 1881.

Animosity between rival gangs of cattle rustlers had been building for months. The rivalry was aggravated by a brother's vendetta. Three months earlier, Farmington vigilantes calling themselves the Farmington Stockmen's Protective Association had gunned down Port Stockton at his cabin near present-day Flora Vista. In revenge, his brother, Ike Stockton, who operated mainly north of the state line in Colorado, rode south and killed one of Port's assassins, Aaron Barker.

Impressed by Ike's daring and looking for refuge across the state line, several of Port's friends came north and fell in with Ike. One such character was Dyson Eskridge—hence the band became known as the Stockton-Eskridge gang. When the Farmington vigilantes rode north to capture the renegades and take them back to New Mexico for punishment, the stage was set.

On Sunday, April 11, some twenty-five to fifty of the armed horsemen approached Durango, headquarters of the Stockton-Eskridge gang, only to find their equivalent, the Durango Committee of Safety, engaged in a lynching. The town was crawling with vigilantes of its own. Exercising a rare degree of prudence, the Farmington bunch skirted the town and continued north three miles to hole up at Animas City.

Just after high noon the next day, they sought a vantage point on the mesa east of Durango. Members of the Stockton-Eskridge crowd got wind of their presence. Scrambling out of the saloons, they charged the hill, six-guns blazing. During the ensuing skirmish shots rained on the town—some reports say as many as a hundred. Flying bullets ploughed up the dirt streets and peppered the buildings. A shot slammed into the West End Hotel just above the head of its proprietor. Another struck the sign in front of the Brunswick Billiard Parlor. One went through the office of the *Durango Record* newspaper, stirring the wrath of its editor.

Bank of the San Juan cashier Alfred P. Camp hid the money and closed the bank. Up on the hill, Peter J. Keegan persuaded the Farmington bunch to hold their fire while his family scurried to the Luttrell house, the only brick dwelling in the neighborhood. Down in town many bystanders didn't know what the fight was about, but a few joined the fray anyway. When the Durango gunfighters saw one of the Farmington riders fly from his saddle, they let out a whoop, unaware that he was only the casualty of a spooked horse. The incident provided a moment of glee for the Durangoans, one of comedy for the New Mexicans on the hill. Eventually the Farmington bunch retreated down the valley, back to New Mexico.

Casualties: Stockton-Eskridge gang—none; Farmington Stockmen's Protective Association—none reported; two bystanders slightly wounded. [1, 2, 7, 13, 181, 195]

gunfighting relatives. The records of the basin's wild frontier days, like those in the larger West, offer examples of family-related gunfighters.

Frank and George Coe, immigrants to the basin from Lincoln County, New Mexico, were cousins, as were Texas gunmen John Wesley Hardin and Emmanuel Clements Jr. Sets of brothers commonly allied as outlaws, sometimes as law enforcement officers. Bob, Emmett, and Grat Dalton; James, Ed, Jim, and Bat Masterson; and Bob, Cole, Jim, and John Younger were blood as well as kindred brothers. In the San Juan Basin brothers Ike and Port Stockton teamed up, as did siblings Dyson and Harg Eskridge. [85, 126, 190, 231]

Hamlet Cabin. At the Aztec Museum Pioneer Village you can see a log structure measuring fourteen by fifteen feet that was moved from its original location four miles up the San Juan River from Farmington. Francis M. Hamlet lived here, and one of his 1880 Christmas Eve party guests was a victim in San Juan County's most celebrated shooting.

Among Hamlet's invited friends were Dr. John W. Brown, who had moved his family from Pueblo the year before, and his son, George Brown. Hamlet did not invite Port Stockton, the area's worst gunslinging miscreant, or Stockton's three sidekicks, Dyson Eskridge, James Garrett, and Oscar Pruett. Three of the Stockton crowd showed up anyway; accounts differ as to whether Stockton was among them.

Outside the cabin, the three troublemakers shot off their pistols and created a row, prompting Hamlet to fling open the door and peer into the darkness for the source of the commotion. When George Brown looked also, a bullet fired by one of the rowdies cut him down and he died on the spot.

Contrary to what you see in the movies, it was customary in the West to disarm yourself before entering a dwelling. The usual practice was to unfasten your belt and loop it, holster and pistol attached, over your saddle horn. An alternative was to hang it near the door or ask the host to take care of it. Having gotten wind that some of the Stockton crowd were bent on disrupting the gathering, the guests had done neither. They were prepared to return fire.

Pruett carried no firearm, and there is no evidence he was involved in the disturbance. But when a reveler inside fired through the door, the bullet pierced Pruett's heart. He died for keeping bad company. His fair-weather friends rode off through the sage under cover of night.

The more-or-less law-abiding element's scorn for Stockton and his ilk is demonstrated by how they treated Pruett's body after it was slung

An 1880 murder at Francis Hamlet's cabin, then located up the San Juan River from Farmington, spurred a series of reprisals and a gunfight at Durango between rival New Mexico and Colorado gangs. The cabin is now part of the Aztec Museum Pioneer Village.

over a horse and taken to Farmington's adobe schoolhouse. No one wanted to bury it. Some accounts say the three culprits outside the Hamlet cabin were Eskridge, Garrett, and Pruett. Others put Stockton at the scene. In either case, the *Durango Record* reported that Stockton was there. Whether he was, in fact, one of the party crashers mattered little to Stockton's enemies. They were more than willing to believe he had a hand in the affair or, at the very least, gave harbor to those who did.

This animosity led to several events: the death of Port Stockton a few weeks later at the hands of the vigilante Farmington Stockmen's Protective Association; a vendetta by Port's brother, Ike; Ike's murder of Aaron Barker; and the ferment of gang rivalry and the basin's most famous gunfight. [13, 51, 85, 107, 192]

Hammond Project. This irrigation system starts with a diversion dam on the San Juan River at Blanco. A hydraulic pump on its main canal lifts diverted water so it can flow to irrigate 3,933 acres stretching twenty miles down the San Juan River to Farmington.

Built by the Bureau of Reclamation in the 1960s, the project is operated by the Hammond Conservancy District. [260, 265]

Harman, Fred (1902-1982). Comic strip characters Red Ryder and his sidekick, Little Beaver, were said to be from Rimrock, Colorado (there is no such place), but they really galloped out of their author's ranch near Pagosa Springs. As a boy in St. Joseph, Missouri, Harman liked to draw with his two brothers at the kitchen table. When a relative sent off one of his pictures to the St. Joseph *News Press,* he became a published artist at age six.

In the 1920s Harman drew alongside Walt Disney. He and Disney started a movie business, but it didn't work out. His first comic strip, *Bronco Peeler and Little Beaver* didn't work either until he replaced Bronco Peeler with Red Ryder. Harman's *Red Ryder* comic strip appeared in 750 newspapers; radio and movie spinoffs made him rich. Red Ryder, often in the person of actor Wild Bill Elliott, rode in over forty films. On the radio he outdistanced the Lone Ranger.

Harman retired Red Ryder and Little Beaver in the 1960s to become one of the West's foremost artists, painting more than 400 oils. After moving to Phoenix, he died of emphysema at the age of eighty.

Many of Harman's paintings are displayed at his studio, now the Fred Harman Art Museum, on U.S. 160 a couple of miles west of Pagosa Springs. [47]

Haviland Lake State Wildlife Area. This body of water covers sixty-five acres at 8,100 feet elevation. The lake is surrounded by quaking aspen and ponderosa pine. The Hermosa Cliffs form a backdrop to the west. Managed by the Colorado Division of Wildlife and the Forest Service, the lake features primarily rainbow trout and a few brook trout.

Forest Service 671 reaches Haviland Lake a half mile after going east off U.S. 550 eighteen miles north of Durango. [250]

Hayden Survey. The best known of four "Great Surveys" by U.S. Geological Service teams that explored the West. It mapped the basin in the late 1870s. Ferdinand V. Hayden's survey normally had six to eight groups in the field. Each included geographers, geologists, naturalists, topographers, and a journalist or photographer. Packers, cooks, and journeymen supported the teams. Among Hayden's photographers was William Henry Jackson. Their records and descriptions gave the general public its first exposure to the mysterious, ancient dwellings of southwestern Colorado. Although the surveyors encircled the Mesa Verde Plateau, they did not investigate its canyons where the now famous sites lay in wait for a later discovery. [69, 228]

Hendrickson, William P. (ca. 1850–?). A settler who pioneered on both sides of the basin's Colorado–New Mexico border. The promise of mineral riches first attracted Hendrickson and his wife, Marion, to the basin. They tried to make a go of it at Animas City but grew weary of prospecting with little reward. In the summer of 1875 Hendrickson and his brother scouted the surrounding countryside for potential farmland. Near the junction of the Animas and San Juan Rivers they found green meadows and the promise of longer growing seasons than they could hope for in Colorado. With a few other families they selected farmsites and, in 1876, became the first permanent settlers on the site Farmington now covers. As was often the case with early settlers, they had no right to be there—not until 1880 would the land be opened for homesteading.

An 1880 incident at Francis M. Hamlet's cabin near Farmington led to an embarrassing confrontation for Hendrickson. That Christmas Eve, while Hamlet's invited guests were reveling at a party inside, Oscar Pruett and two other hooligans created a row outside the cabin. When George Brown came to the door, they shot and killed him. Pruett died from return fire. Hendrickson was returning to Farmington from another Christmas party near Kirtland the next morning when he approached a rider unknown to him. (An unusual happening in the sparsely populated settlement.) The stranger told him of the killings. With his knowledge of the incident, Hendrickson was on the lookout as he got close to Farmington. He saw nothing suspicious but did come upon a funeral where Brown was being put to rest at the cemetery.

But Hendrickson found out that Hamlet's guests had taken Pruett's body to the one-room adobe schoolhouse. For Pruett, like the other rowdies at the cabin, was a henchman of Port Stockton, probably the basin's most abhorred outlaw. No one, not even Pruett's cousin, was brave enough to risk the community's wrath by taking charge of the miscreant's remains. Three days later the body was still in the schoolhouse.

For the sake of dignity, Hendrickson persuaded some respected members of the community to join him and bury the body. But as they ascended the cemetery hill to dig a proper grave, they were followed. As they prepared to scrape away the rocky soil, the pursuer warned them to stop or be shot. As an alternative, he suggested that they hire some gravediggers from a camp down the river. But Hendrickson could find none willing to take the risk.

Returning to town, Hendrickson and his friends found the streets deserted, but inside Markley's general store they met the usual gathering of townspeople warming their hands around the stove. Hendrickson gave them a gentle scolding. Embarrassed by their behavior and sensing security in numbers, all the men present filed out of the store to the graveyard, marked off a plot, and dug a grave. With Hendrickson's persistence, Oscar Pruett's body was properly put to rest. [85, 194]

H

Hermosa, La Plata County, Colorado. In 1881 this stage station welcomed travelers with two hot spring baths, a general store, post office, and flour mill, but not for long. Although it continued as a depot, the coming of the Denver and Rio Grande Railway in 1882 brought machine-powered transportation to the Animas Valley and signaled the station's demise.

Hermosa (elev. 6,645 feet) is ten miles north of Durango on U.S. 550. [101, 248]

Hesperus, La Plata County, Colorado. The Hayden survey party put this name on a nearby mountain, and the title transferred to the community. The village got started in the 1880s to serve coal miners but was never incorporated, though it still has a post office.

Hesperus (elev. 8,000 feet) is eleven miles west of Durango on U.S. 160. [69, 219]

Hesperus Pass. The route of U.S. 160 went this way to get from Durango to Hesperus until a new highway took a more direct route in the early 1950s. The pass is the transition between the La Plata River drainage and Wildcat Canyon. The Rio Grande Southern Railroad crossed the divide to the north of the highway.

Hesperus Pass (elev. 8,019 feet) is west from Durango two miles on U.S. 160, then south six miles on La Plata 141 (Wildcat Canyon Road) and La Plata 125. [219]

Hillerman, Tony (1925–). Journalist, essayist, and mystery writer who sets many of his novels among the Navajos of the basin. Some critics observe that his treatment of central characters, such as Joe Leaphorn and Jim Chee of the Navajo Tribal Police, adds cultural value to Hillerman's novels. The detectives are often caught in conflicts between their Native American heritage and their Anglo education, as are many Navajos in real life. His stories show how many Navajo ways still live in union with the basin's arid landscape.

Born at Sacred Heart, Oklahoma, Hillerman received citations for service in the army during World War II and earned a degree from the University of Oklahoma before becoming a reporter for the *Borger (Texas) News Herald,* a bureau manager for United Press International, and editor of the Santa Fe *New Mexican.* He entered academic life as assistant to the president of the University of New Mexico, where he earned a master's degree and became a professor of journalism. In addition to writing novels, Hillerman contributes to several magazines and journals. He received the Edgar Allen Poe Award from the Mystery

Writers of America for *Dance Hall of the Dead* (New York: Harper's, 1973). [241, 242]

Hinsdale County, Colorado. A county that illustrates the annoyance of many county boundaries. A fraction of this unit rests south of the Continental Divide and within the San Juan Basin. Even though that portion consists entirely of national forest lands and has few, if any, residents, Hinsdale's sheriff has to serve it. And he has to go a roundabout way through three other counties (Mineral, Rio Grande, and Archuleta) to get there on paved roads. The county was named after Colorado lieutenant governor George A. Hinsdale.

Hinsdale County was established in 1874. County seat: Lake City; population in San Juan Basin: 0 (est.); area within basin: 300 square miles (est.). [208]

historic places. Basin sites listed in the National Register of Historic Places include both ancestral Puebloan and early European settlement places. Preservation is a principal motive for registering sites of historical significance. Therefore, where preservation has not begun, the register may omit a precise location. (Some sites, not included here, are identified only by number.) Here is where some registered places are located:

Arboles vicinity. Labo Del Rio Bridge on Archuleta F40 over the Piedra River.

Aztec. Districts: Main Avenue Historic District, bounded by Main Avenue on the east, Chuska Street on the south, the alley between Park and Main Avenues on the west, and Chaco Street on the north; Lovers Lane District, bounded by Rio Grande Avenue on the east, Zia Street on the south, Park Avenue on the west, and U.S. 550 on the north. Structures: H. D. Abrams House, 403 North Church Street; American Hotel, 300 South Main Avenue; Austin-McDonald House, 501 Rio Grande Avenue; Aztec Motor Company Building, 301 South Main; D. C. Ball House, 300 San Juan Avenue; Daws-Keys House, 421 North Church Street; Engleman-Thomas Building, 200 South Main Avenue; Denver and Rio Grande Western Railroad Depot, 314 Rio Grande Avenue; buildings at 202 Park Avenue and 500 White Avenue; Lower Animas Ditch from Church Avenue to Lovers Lane District; Harvey McCoy House, 725 Pioneer Avenue; McCoy-Maddox House, northwest corner of Maddox Avenue and Aztec Boulevard; McGee House, 501 Sabena Street.

Aztec vicinity. Aztec Ruins National Monument.

Blanco vicinity. Frances Canyon Ruin.

Bloomfield vicinity. Halfway House Archaeological Site; Salmon Ruin; Twin Angels Archaeological Site.

Cortez vicinity. Mud Springs Pueblo; Yucca House National Monument.

Dolores. Southern Hotel, 101 South Fifth Street.

Dolores vicinity. Anasazi Archaeological District; Escalante Ruin; Beaver Creek Massacre Site; Hovenweep National Monument.

Dulce. Jicarilla Apache Historic District, bounded by Main Street, New Mexico 17, Apache, Keliiaa, and Sand Hills Drives.

Durango. Places: Main Avenue; Durango and Silverton Narrow Gauge Railroad; East Third Avenue Historic Residential District, extending from Fifth to Fifteenth Streets. Structures: Newman Block, 801–813 Main Avenue; former Colorado-Ute Power Plant at Fourteenth Street and the Animas River.

Durango vicinity. Durango Rock Shelter Archaeological District.

Farmington vicinity. Christmas Tree Ruin; Crow Canyon Archaeological District; Cottonwood Drive Site; Gallegos Wash Archaeological District; Simon Canyon; Hadlock's Crow Canyon; Jaquez Site Ruin; Prieta Mesa Site; Star Rock Refuge.

Largo Canyon. Forty-eight ancestral Puebloan structures among the Pueblitos of Dinetah.

Mancos. Mancos Opera House, 136 West Grand Avenue.

Mancos vicinity. Ute Mountain Ute Mancos Canyon Historic District; Lost Canyon Archaeological District; Mesa Verde National Park.

Pagosa Springs. Chimney Rock Archaeological Site.

Pleasant View. James A. Lancaster Site; Lowry Ruin; Pigge Site.

Rico. William Kauffman House, Silver Street; Rico City Hall, corner of Commercial and Mantz Streets.

Silverton. Silverton Historic District.

Silverton vicinity. Cascade Boy Scout Camp, Lime Creek Road.

Yellow Jacket vicinity. Yellow Jacket Pueblo. [203]

hogan. The shelter of a Navajo family. To its occupants, the home is the center of their world and the hogan is a special religious place. It has a dome-shaped roof of logs covered with dirt. A hole in the center of the roof ventilates smoke from the fire beneath. Some dwellings may have windows, beds, and modern appliances, but traditional hogans have no windows and a flap covers the only entrance. As a gift of the gods, the hogan has a special place in the Navajo's world. It is a sun symbol; its east-facing doorway assures the family within it will see the rising sun. Navajos carry on their native religious ceremonies only in hogans. [82]

hogback. A sharp ridge projecting above the surrounding terrain, like the one north of U.S. 64 between Farmington and Shiprock at Waterflow.

homesteading. While it is true that many pioneers in the San Juan Basin and other parts of the West got 160 acres by making improvements and living on the land for five years, the homestead laws outlined only one of many means by which the public domain was taken for private ownership.

The quest for land, or the control of land, caused an underlying tension throughout the West. Ever since the evolution of the modern state, the granting of property rights has been a policy question for the governors. It has been a means of expanding the power and meeting the objectives of the rulers, whether they be kings or congressmen. Spain, for example, in both the Old World and the New, set land transfer policies so that colonists would become permanent residents of the territory it claimed.

Westward from the original thirteen colonies—and growing largely from the New England experience—came a system of land disbursement based on government surveys that broke the territory into squares of one mile on each side and subdivided into smaller squares. Land grant and sale policies raised revenue for the young republic and encouraged westward expansion. There were many parts to the process. Congress passed, if not hundreds, at least dozens of general laws for land distribution. Here is a sampling:

Military bounty acts: Congress granted land to veterans of the Civil War. This followed the practice of appropriating land, instead of cash, dating back to colonial days.

Preemption acts: Preemption has been described as legalized squatting. To get land this way, you rode into the wilderness and made some "improvements." Then, when the government had surveyed the territory and set the land for sale, you went to the land office before the auction and paid the minimum price of $1.25 per acre.

Timber and Stone Act of 1878: Under this act a settler could buy land for $2.50 an acre. Contrary to the intent of the law, which was to encourage settlement, timber companies used proxies to acquire large acreages.

Land grants to railroads: Congress used the public domain to help and encourage railroad companies expand across the continent by giving them alternate sections within strips of land from six to forty miles wide.

Homestead acts: These enactments provided free land, as much as 160 acres, to settlers who moved onto surveyed land, made improvements, and resided there for five years. The legislation was intended to help would-be farmers, but cattlemen used homesteading to get control of water sources and monopolize the surrounding range.

The list could go on to include desert land reclamation acts, timber culture acts, grants to states, and a myriad of other provisions. Much land was simply sold at auction. These enactments had to be carried out by overburdened land offices, many operating with few precedents to guide them. The scramble for land pitted settlers against railroads, homesteaders against cattlemen, cattlemen against sheepmen, and pioneers against moneyed easterners at the auctions. Bogus entrymen, speculators, and land sharks used fraudulent testimony and hush money to manipulate sometimes corrupt officials and get title to the land. The confusion and conflicts stemming from the quest for land gave rise to much of the discord and violence of the West. Although the romance of the pioneer defying the wilderness and staking a claim is ingrained in the national psyche, it is largely a false notion. [32, 105, 130]

horse. To the Native American, the coming of the horse and his learning to ride it must have been like the Anglos' invention of the automobile—revolutionary. Instead of struggling to move his camp six miles in a day, the horse-equipped Apache could migrate thirty miles. Mounted natives could follow their game for hundreds of miles. And in the basin Utes on horseback found it easier to raid Navajo livestock than to chase after game.

The Apaches became the first native equestrians when they stole horses from Juan de Oñate's Upper Rio Grande paddocks in the early 1600s. The use of horses became widespread as the Apaches traded animals to other tribes. When the Utes and Navajos got horses from the Apaches in the mid-seventeenth century, they became some of the earliest natives to use them for transportation.

When the Spanish fled New Mexico after the Pueblo Revolt of 1680, they left many horses behind. Some fell into the hands of natives, others became feral. Their descendants still roam wild at the northern edge of the basin in Disappointment Valley and on the basin's eastern flank in Carson National Forest. [155, 204]

Hovenweep National Monument. Among the desolate mesas and small canyons north of the San Juan River in Montezuma County, this monument includes four sets of isolated ancestral Puebloan sites near the western edge of the basin: Holly, Hackberry Canyon, Cutthroat Castle, and Goodman Point. Two additional sets are in Utah. An 1854 Mormon expedition first reported the ruins, and William Henry Jackson, who visited the ruins in 1874, first applied its name, which is Ute for "deserted valley." President Calvin Coolidge established the monument in 1923.

The ancestral Puebloans of Hovenweep were culturally related to the inhabitants of Mesa Verde and Aztec Ruins. After living for centuries in scattered villages, they moved from their open valleys and mesa tops in favor of larger settlements at canyon heads to protect their water sources. From these elevations they surveyed the surrounding countryside and used sheet-water irrigation to raise crops. During the latter part of the thirteenth century, drought, failing crops, and perhaps warfare with other natives prompted them to leave and drift south.

The monument straddles the Colorado–Utah border twenty-five miles west of Cortez. [162]

Howardsville, San Juan County, Colorado. The dean of San Juan prospectors, George W. Howard, planted a cabin at this spot in Baker's Park in 1873. Others joined him to build a mining camp worthy of a post office designation a year later.

When the hamlet became the county seat, its future seemed assured. But the town failed to take advantage of its location at the mouth of Cunningham Gulch—the end of the Stony Pass road from Del Norte. When investors sought a smelter site, they found the land overpriced and went to Silverton. The county seat followed. Even so, the town lived as long as the mining did—well into the twentieth century.

Howardsville's remains are four miles east of Silverton on Colorado 110. [96, 97, 115]

Ignacio, La Plata County, Colorado. After the Bureau of Indian Affairs set up its Ute agency here in 1876, trading posts and a community grew up around it. When the Ute Strip was opened for white settlement in 1899, the new arrivals cleared land, built irrigation systems, and established a thriving community. The bureau's Southern Ute Agency and the Southern Ute tribal headquarters are here.

In 1908 two different town promoters, one of which was the Ignacio Townsite and Development Company, bought land from Utes and platted townsites. One developer bought land to the north, the other to the south, with the understanding that each would start selling land where the land joined so the town would mature at one location. That agreement went awry, however, and the town developed in two sections with a vacancy between.

The arrival of Anglos as settlers, merchants, and agency employees amidst Utes living on their allotted lands promoted contact between the two groups. Among the pioneers were many Hispanics. This combined Ute, Anglo, and Hispanic heritage has prompted the townspeople to call their town a "tri-ethnic community."

Ignacio (1990 pop. 720; elev. 6,432 feet) is twenty-five miles southeast of Durango via U.S. 160 and Colorado 172. [187]

Indians. North American natives. These aborigines, like the indigenous people of South America and the West Indies, were named by explorers who assumed those regions were an extension of Asia and the region's people to be of the same race as those of India. Although referred to as Native Americans today, the term "Indian" is unavoidably ingrained in white culture. Standard works use the term as do government agencies (as in "Bureau of Indian Affairs"). Some authors use the term "North American Indian" to distinguish these people from other American natives as well as Asians. By an act of Congress, Native Americans became U.S. citizens in 1924. [74, 118, 220]

Indian reservation. When native tribes were forced to give up their lands, treaties reserved certain portions for their use, hence the term. Of the 300 tracts held for the benefit of Native American tribes, four have territory in the San Juan Basin. There are some 2 million Native Americans in the United States; 800,000 live on or next to reservations.

After the reserves were established in the late nineteenth century, the United States allotted parcels to individual natives under a plan for them to become farmers—the 1895 Allotment Act. (The Weminuche Utes had a land-in-common agreement in place and avoided this program.) The scheme not only failed but also split up many reservations. Lands not allotted to tribal members were opened to white settlement. Such was the nature of the Ute Strip in the Southern Ute Reservation. By 1934, Congress realized the policy was failing, called for the restoration of tribal governments, and began providing aid to tribal groups.

Meanwhile, land ownership within the reservations had been fractured beyond repair and remains in diverse ownership: there are trust lands held by the government for the tribes, nontrust lands owned by the tribes, allotted parcels held in trust, and tracts owned by nonnatives. (When a reservation is said to "encompass" a certain acreage, the area may or may not include nonnative lands). The Southern Ute and Ute Mountain Ute Reservations together stretch along Colorado's southern border from Utah to a point south of Pagosa Springs. The only ones in the entire state of Colorado, they lie wholly within the basin. On the east side of the basin in New Mexico stretches the Jicarilla Apache Reservation; on the west side, the New Mexico portion of the Navajo (the latter extends into Arizona and Utah).

Here are the basin's reservations with their headquarters locations, 1990 populations, and areas:

Jicarilla Apache—Dulce; 2,400; 1,159 sq. mi. (more than four-fifths within the basin).

Navajo—no basin headquarters; 167,000; 25,000 sq. mi. (less than one-fifth within the basin).

Southern Ute—Ignacio; 1,100; 480 sq. mi.

Ute Mountain Ute—Towaoc; 1,500; 922 sq. mi. [217, 220, 249]

The basin attracts more than tourists and new residents; 32,000 enthusiasts registered for the 1996 Ignacio motorcycle roundup. The cyclists spread out to other communities and created this scene in Durango.

To supplement revenues derived from their natural resources, many Native American tribes have turned to gambling casinos, such as this one operated by the Southern Utes at Ignacio.

Indian tribe. The United States recognizes four Native American tribes with reservation territory in the basin: Navajo, Jicarilla Apache, Southern Ute, and Ute Mountain Ute.

Before European settlement, tribes were fluid. Restrained only by natural elements and other tribes, they roamed freely throughout the West. Since they drew no maps, we cannot describe precisely the territory of each group. Intermarriage and crossbreeding further confuse the picture, but we know the San Juan Basin was occupied primarily by Utes in its northern sector; Navajos and intruding Apaches used the southern part. The tribes were often at odds with one another. The Utes and Navajos, traditional enemies, fought for control of Pagosa's hot springs. Utes raided Navajo communities and took livestock. Whites exploited the animosity among tribes. Kit Carson took advantage of long-standing Ute-Navajo enmity by using Utes in the army's war on the Navajos.

What made up a tribe among American aborigines is a subject for debate among anthropologists. "Apache" designates a group of tribes while "Ute" indicates a number of bands. The Navajos call their unit a nation. Alien as all these terms were to the natives, U.S. treaty negotiators needed to deal with an organization—in white men's terms, a "legal entity." As encouraged, or sometimes required, by treaty negotiators and Indian agents, Native Americans formed artificial groups that otherwise might never have evolved. Ancestors of the Southern Ute tribal members, for example, belonged to both the Mouache and Capote bands. [118, 119, 220]

Jackson, William Henry (1843–1942). Among this legendary photographer's most famous pictures are those taken in the San Juan Basin of Utes, ancestral Puebloan sites, and narrow-gauge trains clinging to the side of the Animas River canyon.

As a native of Keeseville, New York, Jackson volunteered for service in the Union army where he sketched camp life and painted portraits of fellow soldiers. After leaving the service and making a transcontinental journey, he began his career in 1868 as a photographer in Omaha. While taking pictures along the Union Pacific Railroad, he met Ferdinand V. Hayden, head of the U.S. Geological Survey of the Territories. Through this acquaintance he become official photographer to the Hayden Survey of the American West and joined a crew that worked in southwestern Colorado. There he found, photographed, and documented pueblo remnants in Mancos Canyon. But in contrast to landscapes and cliff dwellings, he found the Utes unreceptive to imaginative photography unless bribed to pose. They resisted his efforts and sometimes intervened on horseback to prevent his making views.

After spending almost ten years in the field, Jackson set up his Denver photographic and publishing business in 1879. His catalogues boasted a selection of over 10,000 western landscapes in various shapes, sizes, and formats. His sales wagons spread out from Denver to Leadville and other communities. During the next two decades some of his biggest customers were railroads encouraging easterners to use the West as a tourist playground. To build business, railroads commissioned him to enhance their promotions. During this period he was likely the world's leading photographer of railroads, taking perhaps 30,000 pictures of engines, trestles, and tourist hotels, not to mention landscapes tailored to catch the imagination of the city-bound easterner. Although aimed at the commercial market, much of his postsurvey work displays the artistry and imagination he developed through his earlier photography. After two decades in Denver, Jackson moved to Detroit and became part owner of the Detroit Publishing Company, one of the world's largest distributors of colored postcards, prints, and lithographs. After the company collapsed in 1924, Jackson's 40,000 glass plates found their way to the Colorado Heritage Center in Denver.

Jackson, Wyoming, bears his name, as does Jackson Butte near Mesa Verde. [60, 68]

Jackson Gulch Reservoir. See **Mancos Project.**

Jackson Lake State Wildlife Area. A colony of Mormons built a reservoir here in the 1880s and stored irrigation water drawn from the La Plata River. Its fishery now includes bass, catfish, and trout. Waterfowl frequent the lake.

Managed by the New Mexico Department of Game and Fish, the area is four miles north of Farmington on New Mexico 170. [85, 227]

Jicarilla Apache. One of the five Apache tribes, the group has about 2,400 members. The Apaches earned a reputation as the most ferocious of all Native American fighters and produced the fierce warriors Cochise and Geronimo. The Spaniards, using the Zuni word for "enemy," called them *apaches de Nabajo,* or "enemy of the Navajo," which they were, even though both tribes had the same Athapascan ancestors. The Jicarilla reservation lies on the eastern edge of the San Juan Basin, but before white settlers came and the army constricted their territory, they used a much larger region and even hunted the grasslands east of the Continental Divide.

As with other native tribes, the Jicarillas' most successful battle spelled their doom. On March 30, 1854, at Cienequilla in the Embudo Mountains some twenty-five miles south of Taos, over 200 warriors under Chief Chacon entrapped sixty U.S. soldiers. After a three-hour battle only thirty-six cavalrymen lived and made it back to Taos. The defeat prompted retaliation. Using Pueblo Indians as scouts, troops from Fort Union pursued the Jicarillas west across the Rio Grande. This time, Chacon's entrapment scheme failed and his warriors were vanquished. [61, 120, 198, 220, 245]

Jicarilla Apache Reservation. A 1,159-square-mile territory reserved in New Mexico for the Jicarilla Apache Tribe. The area stretches south from Colorado and lies to the east and south of the Carson National Forest within Rio Arriba and Sandoval Counties. A portion laps to the east slope of the Continental Divide.

J

After Chief Chacon's defeat in 1854, a series of skirmishes and aborted agreements led to U.S. domination of the Jicarillas. By 1868, they were conquered, but abandoned by their conquerors—they had no country and no reservation. An executive order in 1874 prescribed a reservation north of the San Juan River from west of present-day Farmington extending east. Two years later, because the Jicarillas refused to move there, President Ulysses S. Grant returned the region to the public domain.

Another order, this one in 1880, set up a reservation in the general area of the north half of today's reservation. Government vacillation reversed this order also, and the Jicarillas were told to join the Mescalero Apaches on their reservation some 300 miles south. That plan didn't work either.

Finally, an executive order by President Grover Cleveland set up a permanent reservation east of present-day Carson National Forest in 1887. Before and after that turning point, the Jicarillas faced periods of abject poverty. For years, pine trees on the reservation bore scars where they cut away the bark to feed off the underlying membrane. Additions since 1887, the last one in 1982, have doubled the reservation's original area to its present size. [61, 120, 198, 220, 245]

Joe Moore Reservoir State Wildlife Area. Turkey Creek irrigation diversion feeds this 45-acre impoundment north of Mancos. It rests at 7,690 feet elevation. A minimum conservation pool assures fish survival. Rainbow trout, largemouth bass, and green sunfish use the lake, which is surrounded by ponderosa pine and aspen woods.

Operated by the Colorado Division of Wildlife, Joe Moore Reservoir is nine miles northwest of Mancos on Colorado 184 and Forest Service 589. [250]

Junction City, San Juan County, New Mexico. The county seat that wasn't. After San Juan County was organized in 1887, the voters took the county seat away from Aztec and gave it to this town across the Animas River from Farmington, on the peninsula between that river and the San Juan—or so they thought. A year after the records were taken to Junction City, however, the territorial court decided Aztec had won the election, so the residents of Aztec rode to Junction City one night, took the records, and rode back home. After its year as an illegitimate county seat, Junction City faded away. [3, 192]

juniper. See **Rocky Mountain juniper.**

Kennebec Pass. Many hikers reach this pass, which separates Junction Creek and the South Fork of Hermosa Creek, by way of the Colorado Trail.

In 1896 Alf Bisch struck gold here with the chance assistance of his reluctant burro. Unlike his other four donkeys, this cantankerous animal refused to let Bisch tie its hobbles. Upon awakening one morning, he found that even the hobbled burros had wandered far from camp and he blamed the footloose jack for leading the others astray. And again the reluctant animal refused to be caught. This was the last straw for Bisch. He corralled the jack in a shallow swale and, in a fit of anger, threw a good-sized rock at its head. The burro ducked, the stone passed harmlessly past and chipped a ledge of rock protruding above a grassy hillside. Upon investigation, Bisch found the vein to be of quartz, generously impregnated with gold. On his location stake he inscribed "The Jackass Lode."

The Jackass Lode turned out to be a small pocket but a rich one. It produced $8,000. Ignoring the temptation to gamble on further prospecting with reluctant burros, Bisch invested in a West Slope farm.

The pass (elev. 11,740 feet) is thirteen miles north of Durango near Forest Road 171 (an extension of La Plata 204). [173, 219]

Kirtland, San Juan County, New Mexico. Before Kirtland there was Olio, established in 1886 and named by the postmaster, perhaps because he viewed the mixture of Mormons, Anglo gentiles, and Hispanics as a hodgepodge. Mormons from nearby Fruitland founded Kirtland in the 1900s. Soon after settling the new community, they dug an irrigation ditch from six miles upstream, where the San Juan River accepts the La Plata.

Kirtland (1990 pop. 3,552) is nine miles west of Farmington on U.S. 64. [3, 85]

Kline, La Plata County, Colorado. This once-flourishing farming community was settled by Mormons who came up from Arizona in the early nineteenth century.

Kline (elev. 6,803 feet) is fifteen miles south of Hesperus on Colorado 140. [48]

Kroeger, Frederick Wilson (1879–1964). The legacy of this cattleman lives on as a Durango hardware store and the brand he registered.

Kroeger made a thirty-day trek from Denver to the basin with his father, Frederick William Kroeger, and mother, Sofia, in 1885. Their destination was the Montezuma Valley, where they sought a homestead irrigated by the Montezuma Valley Irrigation Company. After moving from there to the Animas Valley in 1896, Fred shared a 220-acre ranch with his brother, Henry, until 1919. He then started his own ranch on Fort Lewis Mesa. Somewhere along the line he registered the L-over-U brand. With another brother, John, he bought a supply store in Durango and founded Farmers Supply of Durango, dedicated to "Service to Agriculture," a forerunner of Kroeger's Hardware. His son, Frederick V. Kroeger, continued the mercantile business after Frederick Wilson's death. And the L-over-U brand still adorns cattle grazing on the basin's mesas. [53, 184]

La Baca (La Boca), La Plata County, Colorado. In Spanish boca means "mouth," in this case referring to the mouth of the Pine River as it opened into the San Juan several miles south. (The junction now lies beneath the Navajo Reservoir.) Like many places that are now mere names on a map, this community was a frontier settlement.

La Baca is just north of the Colorado–New Mexico state line on Colorado 172 eight miles south of Ignacio.

L'Amour, Louis (1908–1988). Best-selling author of Western novels who had a place near Hesperus and stayed at the Strater Hotel in Durango. There the Diamond Bell Saloon's tinkling piano inspired his writing.

During the Great Depression Louis Dearborn L'Amour left his North Dakota home to travel and work odd jobs. He wrote poetry and book reviews in the 1930s, but his writing career was delayed while he served in tank destroyer and transportation corps during World War II. After the war he began writing stories for Western pulp magazines. One of his earliest successes, *Hondo* (New York: Fawcett, 1953), became a motion picture, as did many of his later works. In contrast to most of L'Amour's other books, it gained some respect among the critics. Several of his novels followed family sagas—of the Sacketts, the Chantrys, and the Talons. Even though his writing failed to gain the admiration of the literary establishment, he became the first novelist to receive the Congressional Gold Medal. His narrative is fast-paced and straightforward. His works are well researched, surprisingly nonviolent, and reflect his respect for the environment. L'Amour died at his Los Angeles home at age eighty after writing eighty-six Westerns. [11, 224, 242]

L

La Plata, San Juan County, New Mexico. An incident at this place near the Colorado–New Mexico line during the mid-1890s shows a problem sheriffs had before federal agencies reached across territorial and state lines.

One John Scott got into a serious dispute with Bob Caviness, possibly over their claims' boundaries. After a few sharp words, Scott rode to La Plata's general store (the community was known as Pendleton at the time). Caviness followed. When the two resumed their quarrel, Caviness shot and killed Scott.

As a fugitive, Caviness evaded arrest by playing the state line. On the New Mexico side, San Juan County sheriff John Brown was Caviness's friend and always lagged behind in a chase to give Caviness time to cross the line. On the Colorado side, Caviness exchanged fire with La Plata County sheriff Bill Thompson long enough to get back over the line to more friendly territory. Caviness eventually tired of the game and went back to Texas.

La Plata is seventeen miles north of Farmington on New Mexico 170, about five miles south of Colorado. [85]

La Plata City, La Plata County, Colorado. As the principal mining camp in its namesake canyon, this place earned a post office in 1882, only to see it close in 1885, signaling the fragility of the area's mining economy.

La Plata City was up what is now La Plata 124, eight miles north of its intersection with U.S. 160 near Hesperus. [40]

La Plata County, Colorado. Europeans first migrated to this part of the basin from the north as well as from the south. Gold prospectors of the Silverton mining district followed the Animas River down the valley to spend the winter in a more gentle climate. To supply the mining industry, cattlemen and farmers arrived in the 1860s. When the Civil War ceased to be a distraction and the Brunot Agreement reduced the threat of Ute resistance in 1873, settlement quickened.

The Denver and Rio Grande Railway arrived in 1881 and made it to Silverton in 1882, prompting an era of coal mining, ore processing, and related robust commercial activity. The collapse of silver prices in 1893 ended that booming period, although gold and base metal milling continued. Agriculture and tourism also took up some slack. In the 1950s La Plata County rode the crest of the basin's general energy boom when the Atomic Energy Commission reactivated Durango's vanadium mill for uranium processing. Agriculture, tourism, recreation, light manufacturing, education, and general retail activity all play a role in today's economy.

La Plata County was established from parts of Costilla, Conejos, and Lake Counties in 1874. County seat: Durango; population (1990): 32,284; area: 1,691 square miles. [110, 113, 208]

La Plata Mine. Those huge vehicles you see lumbering across the overpass a couple of miles south of the Colorado–New Mexico line on New Mexico 170 are 920-horsepower, 200-ton capacity, bottom-dumping coal trains making the twenty-two-mile trip from this surface mine to the San Juan Generating Station. At the plant, the coal makes electricity for transmission to consumers in New Mexico and Arizona.

The mine's headquarters are a mile east of the overpass on a company road. [146]

La Plata Mountains. The section of the San Juans lying within nine miles of both sides of the La Plata River northeast of Mancos and northwest of Durango. Their highest point is Hermosa Mountain (elev. 13,232 feet). They are a compact group of high peaks named Sierra de la Plata by Spanish explorers who found traces of silver among their ridges and streams. [244]

La Plata River. This stream's importance comes not from its size as much as its inclusion in the Animas–La Plata Project that plans to supplement the river's flow using water from its sister river, the Animas, running at a lower elevation several miles east. The stream starts in the La Plata Mountains on the slopes of Snowstorm Peak north of the former mining camp called Mayday. It then flows fifty-five miles to meet the San Juan at Farmington.

La Posta, La Plata County, Colorado. The Old Spanish Trail passed by this place, but the community wasn't born until the Ute Strip was opened for settlement in 1899. After that it became the only stage stop between Durango and Farmington. Ute subchief Weaselskin lived here; the county road crews still use his name in referring to the nearby bridge over the Animas River. Most of the community's settlers were Hispanic, and their cemetery served Hispanics in nearby settlements. The survivors of the pioneers buried there often built a bonfire the night before a burial, drank strong coffee, feasted, and sang long into the night. They used no coffins, for a dying person often pleaded, "Don't bury me in a box." The settlers buried perhaps thirty of their friends and relatives there. A cedar-post fence edged the graveyard, but there were no stone grave markers. When the community built a chapel a mile north, its churchyard became the new burial ground. The owner of the property

Eighteenth-century Spanish explorers found silver among the rugged peaks of the La Plata Mountains north of Hesperus. They called the mountains Sierra de la Plata, "Saw (Mountains) of the Silver." Courtesy San Juan National Forest.

that included the old cemetery sold the whole plot and the buyer plowed it over for crops.

La Posta is eight miles south of Durango on La Plata 213. [85, 169]

Largo Canyon. This defile running south from the San Juan River near Blanco lives up to its name—largo is Spanish for "long"—for it extends over fifty miles through San Juan, Rio Arriba, and Sandoval Counties into the Jicarilla Apache Reservation east of Counselor. It served as a passageway through the semiarid, southern part of the basin and as a refuge for Puebloans who fled Spanish oppression in Rio Arriba during the seventeenth century. In the gulch's southern reaches, building remains are evidence of how these Puebloans mingled with the Navajos who were already in residence. [153]

larkspur. A flower that looks pretty but is poisonous to cattle and horses (although not to sheep). This perennial beautifies north-sloping aspen forests, seeping springs, and backyard gardens. It is among the first plants to sprout in high meadows, tempting early-pastured cattle. Larkspur poisoning of livestock has no known treatment. [72]

law officer. In the pioneer West it was not unusual for a gunslinger to walk on both sides of the law; he could be an outlaw one day, a lawman the next. And so it was in the basin.

The Animas City Town Council, desperate for an experienced gunman to keep the peace, appointed murderer Port Stockton town marshal, although his term was short-lived. He attacked a barber who had nicked his face, then avoided arrest by leaving town.

Then, as now, a lawman might move from one jurisdiction to another—in the 1880s Robert Dwyer was La Plata County sheriff, then Durango town marshal. One of the West's most famous lawmen, Oklahoman Bill Tilghman, was a deputy sheriff, sheriff, town marshal, deputy U.S. marshal, and Oklahoma City's chief of police. Such a career was possible because he lived seventy years. (The average gunfighter lived to be forty-seven.)

Sometimes law officers quarreled. The Lincoln County War saw Sheriff William Brady side with the Murphy-Dolan ring against Justice of the Peace John B. Wilson and the followers of Alexander McSween. The conflict between Tombstone city marshal Virgil Earp and Sheriff John Behan is the subject of countless Hollywood scripts.

Less glamorous but no less violent was a basin incident involving La Plata County sheriff William J. Thompson and Durango city marshal Jesse Stansel. Thompson was leading a crusade to shut down gambling operations. Stansel, for reasons best known to himself, displayed little enthusiasm for the cause. Their differences led to a cussing match on Main Avenue the evening of January 9, 1906. Witnesses reported that an inebriated Thompson drew his pistol and fired at the marshal. During the ensuing exchange, both lawmen emptied their guns, then Stansel used his pistol to club the sheriff. Four bullets, each potentially fatal, entered or passed through Thompson's body. He died hours later. Marshal Stansel was arrested and charged with murder. A jury found him not guilty.

Many writers of Westerns mistakenly interchange the offices of sheriff, marshal, and U.S. marshal. In fiction a sheriff may answer to the town's mayor even though, by law, sheriffs are elected by the voters of the county. If a sheriff quits or dies, the county commissioners refill the office. In fact, law officers' duties did sometimes overlap. A town marshal might ride out of town after a bandit—with or without the authority to do so. And a sheriff, as did Pat Garrett, might double as a U.S. deputy marshal to supplement his meager county income. The duties of these lawmen are substantially the same today except that most powers

William J. Thompson, sheriff of La Plata County. During a 1906 cussing match about enforcing gambling laws, Durango city marshal Jesse Stansel shot and killed Sheriff Thompson on Durango's Main Avenue. At trial, Stansel was found not guilty. Courtesy La Plata County Historical Society.

previously belonging to justices of the peace have been transferred to municipal or county courts.

U.S. marshal—an officer appointed by the president to serve a federal district court. Because most criminal laws were state or territorial rather than federal, U.S. marshals and their field deputies normally did little day-to-day peacekeeping and were in few gunfights.

Sheriff—the chief law enforcement officer of a county. Sheriffs were elected in both territories and in states but frequently resigned or were killed, creating vacancies that were filled by appointees of the county board of commissioners.

Town marshal or marshal—a peace officer appointed by city or town officials, normally the town council. The overburdened sheriff's inability to subdue the criminal element often prompted the townspeople to incorporate a town so they could raise money and hire a marshal.

Justice of the peace—a magistrate with the power to try minor cases and perform other legal duties within a precinct of the county. The J.P. was often the first judicial contact by an arrested person. He decided whether the accused should be held for trial.

Constable—the officer of the Justice of the Peace Court.

Coroner—an elected official responsible for determining the cause and circumstances of unnatural deaths. [3, 14, 25, 26, 84, 88, 94, 195, 231]

leafy spurge. The roots of this creeping perennial weed, introduced from Europe, can go down thirty feet. Its sap is a milky latex damaging to eyes and skin. Because of its sprouting root system, it is extremely difficult to control. When it gets established, it can exclude all other vegetation and render land useless for cattle grazing. Fortunately, however, sheep and goats relish it. [38]

Lebanon, Montezuma County, Colorado. This locality was called Hardscrabble until 1909, when the Colorado Land and Improvement Company decided it should have a less onerous moniker. The speculators laid out a townsite, built a hotel, cleared land, planted orchards, strung telephone lines, and boosted the place as great fruit country. They were right about the fruit-growing possibilities—the area later proved that—but the development scheme didn't work and the outfit went bankrupt.

Lebanon is seven miles north of Cortez on Montezuma 25. [50]

Lemon Reservoir. See Florida Project.

Lewis, Montezuma County, Colorado. A farming community that got started in 1906 when settlers came to take advantage of water from the Montezuma Valley Irrigation Company.

Lewis is eleven miles northwest of Cortez on U.S. 666. [50]

Lincoln County War. A training ground for San Juan Basin gunfighters. Some combatants migrated north from this bloody conflict in southern New Mexico to make the basin their stomping ground. It is hard to find a conflict of the American West that has been the subject of more writings than the Lincoln County War of 1877–1878. Suffice to say it was a conundrum of merchants, bankers, lawyers, rustlers, lawmen, and hired guns in opposing camps seeking to control the cattle trade and other mercantile activities in that part of the territory. The clash continued until some of the principal characters died in battle and money to pay the hired guns dried up.

Among those who rode north with their families from Lincoln County were George Coe, who fought with Henry "Billy the Kid" McCarty, and George's cousin, Frank Coe. After his arrival in October 1878, George Coe rented two ranches where Aztec is located. Ike Stockton and his reckless brother, Port, soon followed. Both were known as cattle rustlers and desperados, and they sustained those reputations after they came to the basin. Ike became notorious as a leader of the Stockton-Eskridge gang, which terrorized the countryside from Rico to Farmington. [37, 94, 223]

Lizard Head Pass. At the peak of the mining boom in the 1890s, this pass was the scene of a mining camp and the Rio Grande Southern Railroad route. A quarter-mile-long snow shed protected the RGS tracks from heavy snows near the summit, but it didn't prevent ten passengers from being stranded in 1944. They had to be rescued by dog sled. The pass was used by Native Americans, trappers, and packers before it became a miners' wagon route in the 1870s and a railroad way in the 1890s.

Lizard Head Pass (elev. 10,250 feet) is fifty miles northeast of Dolores on Colorado 145. [219]

L

Lizard Head Wilderness. Fifty square miles of this area were first set aside by the secretary of agriculture in 1932 as the Wilson Mountains Primitive Area. Congress enlarged it to a sixty-five-square-mile wilderness in 1980. It straddles the San Juan and Uncompahgre National Forest boundaries at the north edge of the basin. The basin's half of the wilderness features two of Colorado's "fourteeners"—Mount Wilson (elev. 14,246 feet) and El Diente Peak (elev. 14,159 feet). Several other spires exceed 13,000 feet. As the crow flies, the center of the area is nine miles north of Rico. [158]

Lone Dome State Wildlife Area. This 6,000-acre area, managed by the Colorado Division of Wildlife and the Forest Service, features large and small game hunting and Dolores River fishing. Lone Dome is five miles east of Cahone via Montezuma R, Montezuma 16, and Montezuma S. [250]

longhorn. For reasons best explained by a geneticist, when various strains of Spanish cattle got together in what is now Texas, they brought forth beasts with horns gently curling out to vicious points. This is the longhorn, an enduring symbol of the mythic West.

One of the strains came to Hispaniola with Columbus in 1493 on his second voyage. In 1521 Gregorio de Villalobos delivered some of the breed's offspring to the mainland. When the Spanish moved north in the 1690s to set up missions in East Texas, they brought cattle with them. A quarter-century later two Frenchmen exploring the region reported thousands of cattle roaming around a mission on the Neches River.

"American" stock from the east infused these Spanish "black" cattle before the black strain bred with the "Mexican" cattle (also of Spanish lineage) that had moved north into central Texas. Although "longhorn" can refer to any cow with long horns, the rangy animals this mixing produced, some with horns spreading up to nine feet, became known as the longhorn breed. Reflecting their origin, they are also known as "Texas longhorns."

The Civil War interrupted cattle exporting and enticed herdsmen to become soldiers. As the generals fought to Appomattox, the feral longhorns multiplied until perhaps six million were roaming the hills and hiding in the brush of South Texas. Returning veterans found a product ready for sale to army posts, Indian agents, and a meat-hungry nation. Cattle exporting resumed on a massive scale as the Texans drove hundreds of herds numbering thousands of animals as far as California and Canada with stops at intermediate markets and railheads.

But the longhorn's time was short. Cattlemen breeding better stock feared the ticks and fever brought by the wild cattle. They strung barbed wire to defend ranges, secure water holes, and protect their herds from longhorn tick infestation and crossbreeding. Before the twentieth century arrived, the longhorn's heyday had passed. [43, 108, 204]

Los Pinos River. See **Pine River.**

Lowry Pueblo Ruins. Built by ancestral Puebloans about 1060, these ruins had living quarters and ceremonial rooms, called kivas, for about one hundred residents. Controlled by the Bureau of Land Management, the site is nine miles west of Pleasant View on Montezuma CC. [203]

Lumberton, Rio Arriba County, New Mexico. Around 1890 Emmet Wirt brought his cattle and started a trading post somewhere east of the present-day Jicarilla Apache Reservation on the route of the Denver and Rio Grande Railway. The settlement, a temporary home for sawmill workers with accompanying saloons, was called Amargo. Here Wirt started a butcher shop and general store.

After the store lent the post a sense of permanence, a land shark observed that the occupants were squatting with no property rights—a common practice for the time—and filed a homestead on the ground for the purpose of getting title and selling the land at exorbitant prices. A situation that might have led to bloodshed was settled when Wirt persuaded everybody at the post to simply pick up and move a few miles west. The new location became Lumberton, so named in view of the lumber operations underway in the vicinity.

Lumberton is three miles east of Dulce on U.S. 64. [198]

lynching. An execution without due process, like the Silverton hangings of Kid Thomas and Bert Wilkinson for the 1881 murder of town marshal D. C. "Clate" Ogsbury; and the Durango hanging of Henry Read Moorman for shooting James K. Prindle at the Coliseum saloon the same year. Hanging was the most common method used, but the term applies to any form of death prompted by a body of citizens who pass judgment without legal authority. Lynching was not unique to the West, of course. African Americans were lynched in the South long after the West was tamed. Nor did it start on the frontier, although it is apparently of American origin. The term may have come from one or both of a couple of Virginians who took it upon themselves to hang loyalists during the American Revolution—Colonel Charles Lynch and Captain William Lynch (no relation). [89, 188]

Mancos, Montezuma County, Colorado. Like many communities in the upper San Juan Basin, this town started in the era of cattle and homesteading. It was incorporated in 1894, twenty years after discouraged prospectors established ranches in the valley. The Rio Grande Southern Railroad promised commercial stability when it came through in 1891. In 1915 Willa Cather described the town as a place with friendly people, tree-lined streets, sage and paintbrush yards, and fragrant sweet clover surrounded by golden fields of wheat. Through the early 1950s, its commerce was sustained by the Rio Grande Southern Railroad. Named after the river running through it, the community is home for workers who commute to Cortez and Durango.

Mancos (1990 pop. 842, elev. 6,993 feet) is sixteen miles east of Cortez on U.S. 160. [50, 199]

Mancos Project. The West Mancos River feeds 220-acre Mancos Reservoir through a canal in this Bureau of Reclamation project. Built in the 1940s, it is managed by the Mancos Water Conservancy District. The body of water is also called Jackson Gulch Reservoir, a name best not used to avoid confusion with Jackson Lake near Farmington. The reservoir releases water into an irrigation system built by private interests in the late nineteenth century and provides domestic water for Mesa Verde National Park and Mancos.

The reservoir is four miles north of Mancos via Montezuma 42. [261]

Mancos River. Originally called the San Lazaro River, this stream was renamed by the Domínguez and Escalante explorers after one of them broke his hand; *mancos* is Spanish for "crippled." It commences on the slopes of Mount Hesperus, then flows past its namesake town and on to the San Juan sixty-five miles downstream near Four Corners. [104]

Mancos State Recreation Area. A recreation area managed by the Colorado Division of Parks and Recreation and located at Mancos Reservoir (also called Jackson Gulch Reservoir), four miles north of Mancos via Montezuma 42.

Manifest Destiny! As a slogan for American boosterism, this motto purported to justify the nation's theft of other countries' territories. (Historians might say it was a philosophy to justify the expansionist manifestation of nationalism.) The phrase proclaimed territorial expansion to be not just a triumph for liberty but inevitable, a national duty, an American right! The doctrine was without rational foundation. It was dreamed up by John O'Sullivan, a New York newspaperman whose exuberant prose merely played on the aims of expansionists. And just as Patrick Henry's cry of "Give me liberty or give me death!" did not necessarily reflect the feelings of most colonists, so "Manifest Destiny" did not necessarily express the opinion of most Americans.

In the end, this egotistical idea fell short of its claim that Canada, Mexico, Cuba, and the rest of North America belonged to the United States by the dictates of providence. But whether the creed merely reflected an ambition or promoted a cause, it was part of the nation's expansion—a rallying cry for territorial conquest. One of the regions taken (from Mexico) under this doctrine included the San Juan Basin. [28, 130]

Manuelito (ca. 1818–1893). The Navajo headsman who, more than any other, advocated war with the United States and held out the longest against surrender and confinement at Bosque Redondo. At one point he forced the army to abandon Fort Defiance, but new army troops from the east soon retook the area. Recognizing the European immigrants' numerical superiority, however, he led an 1876 delegation of Navajo leaders to Washington, D.C., where he pleaded with President Ulysses S. Grant for more grazing rights and persuaded federal officials to expand the reservation.

With his Mexican wife, Juanita, he made his home near Tohatchi, south of Sheep Springs, New Mexico. While many Navajos were herded to the camp near Fort Sumner, Manuelito's band held out. Having gone to Bosque Redondo with a contingent of prisoners, he recognized the place for what it was, a concentration camp. He fled the compound and headed for the Zuni Mountains far to the west. It was not until September 1866 that he and his twenty-three surviving followers surrendered at Fort Wingate and straggled to the Bosque. Ten years later, with his fellow tribesman, he was permitted to leave the Bosque and live on the reservation. In 1885, realizing the need for the white man's education among his people, he sent his son, Manuelito Segundo, and a nephew

Marmot. This rodent is plentiful at high elevations, where it has a choice of vegetation for food. Sometimes called a "whistle pig" for its shrill cry, in the East it is also called a groundhog or woodchuck. Courtesy San Juan National Forest.

to the Indian School at Carlisle, Pennsylvania. Illness prompted his son to come back to New Mexico where he died shortly after returning.

Manuelito had been the first leader to encourage Navajos to send their children away to schools. But after his son's death, he saw the dilemma of trying to adapt to the white man's ways and sought solace from the bottle. He spent his last ten years fighting alcoholism. Weakened by his addiction, he succumbed to measles and pneumonia at age seventy-five. [64, 82]

marmot. These robust rodents are sometimes called "whistle pigs" due to their shrill call. Although they are known to eat farm crops at lower elevations, in the basin they prefer to live high in the mountains surrounded by succulent vegetation. Their blunt snouts, coarse fur, and bushy tails present an amusing appearance. Where you see one, you are likely to see more, for they dwell in communities. Most marmots in the basin are the reddish brown, yellow-bellied variety; they grow up to two feet long. In the East they may also be called groundhogs or woodchucks. [135]

marshal. An appointed law enforcement officer. Town marshals on the frontier (as now) were appointed by town councils. Citizens tired of rowdiness beyond the control of the sheriff often incorporated a town so they could raise money and hire a marshal. Federal marshals serve and execute orders of federal courts; most of their duties are carried out by deputies.

Marvel, La Plata County, Colorado. Mormon settlers around Kline who built a flour mill here decided to have their own town in 1915.

Marvel is on Colorado 140 fifteen miles south of the U.S. 160 intersection at Hesperus. [48]

maverick. When Texan Samuel Maverick's unbranded herd mixed with neighboring cattle on the open range, he rounded up and claimed all unmarked animals. He was merely following a practice of his fellow cattlemen, but the extent of his indulgence brought his name to mean "unbranded animal." The verb adaptation, "mavericking," is the act of indiscriminately claiming and branding cattle. It sounds less onerous than "rustling," a euphemism for stealing. [43, 54, 108]

Mayday, La Plata County, Colorado. Named after a nearby mine, this community was at the end of a Rio Grande Southern Railroad spur that started taking ore out of La Plata Canyon's mines in the 1890s.

Mayday is four miles north up La Plata 124 from where it leaves U.S. 160 near Hesperus. [175]

McCarty, Henry (1859–1881). Better known as "Billy the Kid." Other names: William Bonney, Henry Atrim, Kid Atrim, William Atrim. Enough books have been written about this outlaw to fill a good-sized library. Let us just say he was a notorious killer whose career has been greatly exaggerated. He was a gunslinger for less than four years, and only about a half dozen of his alleged two dozen killings can be verified. His life is pertinent here only because he was a sidekick or acquaintance of some exiles of southern New Mexico's Lincoln County War who migrated to the basin: George Coe, Frank Coe, Ike Stockton, Port Stockton, Cal Brown. Although the *Historical Encyclopedia of Colorado* mentions that he was a frequent visitor to Durango, I have found no verification of this claim. If McCarty ever prowled the San Juan Basin, he apparently behaved himself. [208]

McElmo Canyon. A gulch on the west edge of Montezuma County that extends into Utah. The site of a 1911 basin oil strike, it is named after a white pioneer prospector who stopped at a spring in the canyon to get a drink and died there of tuberculosis. His partner buried him in the canyon. [50]

McPhee Reservoir. See Dolores Project.

Mears, Otto (1840–1931). Road builder, industrialist, packer, freighter, treaty negotiator, politician, erector of railroads. His ashes are fittingly part of Engineer Mountain near Silverton, for no one did more to ease the basin's development.

A Russian immigrant with a British father, Mears made his way via the Isthmus of Panama to the West Coast, where he joined the California Volunteers. As a member of the First Regiment during the Civil War, he helped keep New Mexico under Union control and then served with Colonel Kit Carson as Carson carried out his scorched-earth strategy against the Navajos.

With his army mustering-out pay, Mears opened a store in southeastern Colorado's San Luis Valley, built a lumber mill, and took up farming. In order to get his wheat to a mill in the Arkansas Valley, he needed a road. So he built one—the first of some 450 miles of toll roads he would scratch and blast out of the Colorado landscape. Many would provide the beds for future railroads (most of which he would build) and for highways still used today. His road building gained momentum in the San Luis and Arkansas Valleys before he pierced the San Juan Basin in the 1870s and 1880s with routes from Lake City to Animas Forks and from Ouray over Red Mountain to Silverton. He built roads from Silverton up the Animas River to Mineral Point and from Durango to Fort Lewis.

With no formal training as an engineer, Mears took a common-sense approach to route planning. He liked to climb the highest peak he could find, take a look around, then start giving instructions. During construction and after the toll roads were in operation, the slender and wiry Mears supervised his operations by dashing about the country in his mule-team rig—sometimes covering a hundred miles a day—cutting a curious figure with his short black beard and little Russian cap.

For this experienced road builder, railroading proved an obvious sequel. From Silverton he ran the Silverton Railroad along his roadbed to Red Mountain and the Silverton Northern along his route to Animas Forks. From Durango he built the Rio Grande Southern over the tortuous route to Rico and Lizard Head Pass. After losing control of the RGS in the panic of 1893, Mears took a recess from the basin and went east to build a line from Washington, D.C., to Chesapeake Bay. With the Mack brothers, he developed the Mack truck.

With such personages as Kit Carson and Felix Brunot, Mears helped negotiate treaties removing the Utes from territory coveted by miners and other white settlers. Helping shape the new state of Colorado, he was a state legislator and one of the state's first presidential electors. He also helped supervise the building of Colorado's gold-domed capitol. During his advanced years Mears held on to his investments in the basin's mining industry. In 1914 he moved to Pasadena, California, where he lived until his death at age ninety-one.

In the solid rock cliff at the edge of Bear Creek Falls south of Ouray, you can read a granite-slab memorial that commemorates his most dramatic feat, the Ouray-Silverton highway: "In Honor of Otto Mears, Pathfinder of the San Juan." [196]

Meeker Massacre. An incident that ended 200 years of Ute efforts to peacefully coexist with Anglo-Americans and sent a spasm of fear through the San Juan Basin. Perhaps better called the Thornburgh Battle, it is also known as the Battle of Mill Creek or the White River Massacre.

As head of the northern Colorado White River Ute Agency in 1879, political appointee Nathan Meeker ineptly pursued the Department of the Interior's agenda for making native hunters and gatherers become farmers. So intent on working his will was he that he tried to get the army to enforce it. Animosity grew as the agent directed the Utes to dig up meadows for crops.

When a half-breed medicine man named Johnson threatened Meeker for plowing under his horse pasture, Meeker panicked. He asked Colorado Governor Frederick Pitkin to consult with federal authorities and bring in the army. A flurry of telegrams convinced officials in Washington, D.C., to authorize military action, violating an agreement to keep the army away from the Ute agency. At Fort Fred Steele, near Rawlins, Wyoming, Major Thomas T. Thornburgh received orders to help Meeker and marched south with 175 mounted troops. As the troops approached the Indian agency, Meeker had second thoughts and urged Thornburgh to halt and join a parley. Thornburgh agreed, but let 120 cavalrymen cross Mill Creek and advance into the reservation anyway. Chief Black of the Utes faced them with 100 warriors. For a few minutes the opposing forces hesitated, apparently ready to reconcile and avoid combat. Then a warrior, Ute or soldier, either deliberately or by mistaking a signal, fired a shot and touched off the Thornburgh Battle.

Major Thornburgh died from a slug behind his ear, but Captain J. Scott Payne took command and rallied the troops. For several days the Utes pinned down the soldiers until forces of the Ninth Cavalry arrived to bolster the besieged troopers. On October 5 they overpowered the Utes. While their warriors held off the cavalry, other Utes vented their anger by slaughtering Meeker and ten other men. They also kidnapped five females, including Meeker's wife and daughter. Through the mouth of Meeker's body the Utes drove a stake "to stop his infernal lying."

The incident prompted demands for military protection. To defend settlers and townspeople in the Animas Valley, troops marched west from Fort Lewis near Pagosa Springs and bivouacked across the Animas River from Animas City. The garrison moved farther west the next year, 1880, and set up Fort Lewis as a permanent station on the La Plata River south of Hesperus. [86, 123, 202]

mesa. A tableland that is higher than its surroundings, often but not necessarily with precipitous slopes on its sides. The term is Spanish for "table." The most famous such landform in the basin is Mesa Verde.

Mesa Verde National Park. This is the only national park in the United States dedicated to the works of prehistoric man—the ancestral Puebloans. A World Heritage Site, it rests in the high plateau country of the basin.It holds within its eighty-one square miles more than 4,000 archaeological sites, including 600 cliff dwellings. The Puebloans built the structures between 600 and 1300 when their civilization flourished. Then they left. Although archaeologists have developed many theories, why they abandoned their centuries-old habitat remains a mystery.

Stories of ancient dwellings circulated among Spanish explorers more than a century before Anglos took up permanent residence in the basin. Prospectors too told of strange cliff cities. William Henry Jackson photographed some of the mesa's sites in 1874. William Holmes of the Hayden Survey examined Mesa Verde cliff houses the next year. By 1879, enough relic collectors knew about the place to prompt John Routt, Colorado's first state governor, to recommend site preservation. But it was not until 1888 that some of the more dramatic and now most famous remnants of Mesa Verde were discovered. Although no one documented exactly who first saw them and a precise date, the most popular version of the discovery of the Cliff Palace site says two cowboys came upon them while searching the chaparral for strays. One can only imagine the awe of Charlie Mason and Richard Wetherill when their eyes fell upon the ancient apartments across the canyon.

Mason and members of the Wetherill family explored the area and gathered artifacts. As others learned of their discoveries, looting accelerated. Selling native relics proved more lucrative than raising cattle or farming. Realizing the significance of what his family was collecting, B. K. Wetherill contacted the Smithsonian Institution, but it had no funds to take advantage of Wetherill's offer to sell much of his collection. Some who grasped the value of the Puebloan heritage made scattered efforts to get the federal government involved in preservation. Others favored state action. Conservation efforts prevailed after Virginia McClurg got the support of the Colorado Association of Women's Clubs and the women successfully lobbied their congressmen. After all the controversies ran their course, Congress established Mesa Verde National Park in 1906.

Although set aside primarily to preserve prehistoric antiquities for study and observation, the park offers unique scenery and varied wild plants and animals. Altitudes in the park vary from 6,000 to 8,500 feet, and several places in the park offer views of the entire Four Corners region; from Park Point you can see into Utah, Arizona, and New Mexico.

In the words of one historian, "Mesa Verde is not, as many people think, an inconveniently located museum. It is the story of an early race, of the social and religious life of a people indigenous to that soil and its splendors. It is the human expression of that land of sharp contours, brutal contrasts, glorious color and light. The human consciousness, as we know it today, dwelt there, and a feeling for beauty and order was certainly not absent."

The park's entrance is on U.S. 160 ten miles east of Cortez, midway between that city and Mancos. Its visitor center is fifteen miles from the entrance. [112, 163]

Mexico. The country in which the basin was located before 1848. Mexico is an Aztec word meaning "place of the moon." The basin left Mexico's jurisdiction at the conclusion of Mexico's war with the United States. By the Treaty of Guadalupe Hidalgo, Mexico gave up its claim to much of what is now the American West, including the part we call the San Juan Basin.

Milagro Co-Generation Facility. This $50 million "co-gen" plant produces steam to purify natural gas at the same time it generates electricity for sale to various utilities. Capable of producing 62 megawatts of power—enough to supply a city the size of Farmington—the operation went on line in 1996. The facility is northeast of Bloomfield. [189]

Million Dollar Highway. U.S. 550 running north from Silverton over Red Mountain Pass to Ouray. The origin of the name is obscure. Some say it was so named because Otto Mears spent a lot of money building it. Others say it has a "million-dollar" view.

Mineral County, Colorado. Just a small part of this county lies southwest of the Continental Divide and, therefore, within the San Juan Basin. Its basin portion lies entirely within the San Juan National Forest.

Mineral County was established in 1893. County seat: Creede; population in San Juan Basin: 0 (est.); area within basin: 200 square miles (est.). [208]

mining. Today the basin's huge open-pit coal mines produce coal by the millions of tons to feed generating stations. Since the beginning of white settlement, coal mining has been essential to the basin's development. It fired the trains, generators, and smelters that made other mining possible. In the 1950s uranium mining brought a flurry of economic activity to the basin. But the practical benefits of coal and uranium fall short of the mystique of gold and silver.

The Spaniards knew precious metals were there. That's why they named some of the mountains Sierra de la Plata (*plata* means "silver"). Perhaps this hint of easy wealth attracted Charles Baker to lead prospectors into the San Juans in 1860. In the 1870s prospectors poured into the region, and mining camps sprang up by the dozens. The persistent hammer of stamping mills soon pervaded the valleys. Smelters fired up to process the ore. Railroads fingered out from Silverton and Durango to carry ore to the smelters, coal to the mines. By 1891 some 500 mines were listed as operating in southwestern Colorado. Among the dozens of mines in the basin, one became eminent after Otto Mears took it over in 1912. Perhaps its name contributed to its fame. It was called the Gold King.

But the San Juans' showcase mine was the Sunnyside. It grew from a claim by George Howard, a member of the Baker party, to become the most productive gold mine in Colorado. Like most prospectors, Howard lacked the capital and business acumen to develop the mine and live through the usual start-up losses. The claim went though part-ownership schemes, mortgages, foreclosures, shut-downs, and start-ups before John H. Terry got complete control in 1900. Through his business sense and intuitive discovery of a rich gold vein, the workings became the San Juan's premier mine. By 1904 it worked a force of 180 to bring out fifteen tons of high-grade concentrate daily. Its prosperity spread to nearby Eureka.

During World War I, the Sunnyside took advantage of rising base metal prices and mined lead, zinc, and copper. But the enterprise rode an economic roller-coaster. When the war ended and prices plunged in 1920, the mine closed, only to reopen again the next year. During the Great Depression the operation shut down again, then resumed production when metal prices surged in 1937. In 1978, 125 men were working in the mine, bringing rich gold ore from a vein below Lake Emma, a body of water formed by an ancient glacier. The lake broke through. Millions of tons of water and mud crashed through the mine, turning its entrance into a monstrous gargoyle and bewildering a guard on duty outside. The date was June 4, 1978, a Sunday. No crews were working, so the deluge injured no one. The Sunnyside survived that disaster, the vagaries of metal prices, and ever more stringent government regulation until 1994. Then it closed for the last time.

The romance of gold mining lingers, fostered by gold-panning exhibitions and old mine tours. But it hides the story of a business that once worked men seven days a week to make a few investors rich. Miners denuded forests. Mills and smelters threw dust and smoke into the clear mountain air. Mines filtered silt and chemicals into crystal streams. Their leftover tailings still do. [10, 55, 96, 97, 111, 113, 115]

Molas Pass. In the late 1980s this place was reported to have the cleanest air in Colorado. That explains why its 140-mile views are so clear. Molas Pass (elev. 10,910 feet) is seven miles south of Silverton on U.S. 550. [219]

Monero, Rio Arriba County, New Mexico. Through this place passed many who molded the basin's history. Domínguez and Escalante were among the Spanish explorers who traversed the Old Spanish Trail through present-day Monero, using a swale six miles east of the place to get over the Continental Divide. Kit Carson used that path, and the Denver and Rio Grande Railway took advantage of the same depression. After Anglo pioneers came, lumber companies moved their timber through here on the way to Chama for milling and shipping to markets.

Monero is thirteen miles east of Dulce on U.S. 64.

Montezuma County, Colorado. Located in the southwest corner of Colorado, Puebloans and then Utes and other Native Americans used its land before it was settled by Europeans in the 1870s. As in other parts of the basin, many early settlers were cattlemen. Sheepmen and homesteaders followed.

When the Montezuma Valley Irrigation Company brought irrigation water from the Dolores River to arable land in the late 1880s, fruit production became an important industry. McElmo Canyon peaches brought awards from the 1904 St. Louis World's Fair. Lebanon area farmers boasted superb apple crops. The Rio Grande Southern Railroad came through in the 1890s to support the area's farming, lumber, and mining operations through the 1940s.

In 1911 test-drilling companies struck oil south of Cortez and in McElmo Canyon to the west. America's demand for energy spurred oil drilling and pipeline building in the 1950s. Uranium prospecting and mining also impacted the economy. Construction of McPhee Reservoir provided jobs for the county in the 1980s; the project's expanded irrigation-enhanced agriculture. From both irrigated and dryland fields, pinto bean production and processing has stabilized the economy. And through good times and bad, tourists have continued their relentless journey through the county on their way to Mesa Verde National Park and other attractions.

Montezuma County was split from La Plata County and established in 1889. County seat: Cortez; population (1990): 18,672; area: 2,097 square miles. [31, 50, 182, 208]

Moorman, Henry Read (?–1881). Victim of a lynching by Durango vigilantes. For reasons unknown but within full view of the patrons at a notorious place of amusement, the Coliseum saloon, Moorman shot and killed one James K. Pringle in cold blood on the night of April 10–11, 1881. Witnesses immediately apprehended Moorman and turned him over to the Durango Committee of Safety, a vigilante bunch. The next evening two signal shots alerted the committee members to gather at a large ponderosa pine in front of the post office. When all were assembled, the culprit was summarily hanged.

The *Durango Record* described the aftermath: "The slight wind swayed the body to and fro! The pale moonlight glimmering through the rifted clouds clothed the ghastly face with a ghastlier pallor! The somber shadows of the massive foliage seemed blacker than the weeds of mourning; and the shuffling of hurrying feet in the dusty road, mingling with the weird whistling of the breeze through the pine boughs, broke upon the ear with a sepulchral tone." [13]

Morgan Lake. This reservoir's basic purpose is to supply water for the Four Corners Generating Station, but it also provides wildlife habitat as well as opportunities for bird-watching, boating, fishing, and windsurfing. Volunteers and station personnel have planted cottonwood and willow sprigs on the lake's shoreline and erected duck and swallow boxes around the lake. Warmwater fish—largemouth bass, channel catfish, bluegill, carp—live in the lake. As a body of open (unfrozen) water in a dry country, the lake attracts water birds: loons, grebes, herons, pelicans, cormorants, ducks, rails, gulls, and ospreys. Its shores, thickets, and riparian woodlands also bring shorebirds and perching birds. In all, more than 200 species visit the lake, many making it their seasonal home.

Located on the Navajo Reservation, Morgan Lake gets its water from the San Juan River. It is ten miles west from Farmington on U.S. 64 and seven miles south on San Juan 6675. [144]

Mormons. It is safe to say that no religious sect had more influence during the white settlement of the intermountain West than did the members of the Church of Jesus Christ of Latter-day Saints. And their influence stretched heavily into the basin.

Persecuted in Illinois, Ohio, and Missouri, Mormons migrated west during 1846–1847 into present-day Utah, then part of Mexico. When the region came under U.S. control in 1848, they sought to establish their own state of Deseret. But the Mormons acquired direct political influence over only Utah Territory, where their shadow government enacted laws to be rubber-stamped by members holding civil offices.

As their colonizing efforts continued, the church leaders apparently had ambitious plans for the basin, as evidenced by Brigham Young Jr.'s interest in the area. He encouraged Luther Burnham to move from St. Johns, Arizona, to the lower Animas Valley and preside over a new Mormon ward. For a time, Young himself kept a home in the basin for one of his wives. In the 1880s and 1890s, Mormons settled in the basin's La Plata River valley, where they established Kline, Marvel, Redmesa, and Pendleton (now La Plata). Jackson Lake, a few miles north of Farmington, was originally built by Mormon colonists. In the Animas Valley they started Kirtland, named for their Ohio settlement, and Fruitland.

Whether the Mormons came to seek opportunities or to proselytize is a matter for conjecture. Regardless of their motive, they played an important role in the basin's maturation. Some took charge of Navajo boarding schools. John Bloomfield secured $10,000 from the church to help settlers south of Aztec pay for an irrigation system. The community now bears his name. [85, 90]

motto. Few jurisdictions smaller than states have mottoes, so even if the basin were a political unit, we can assume it wouldn't have one. If it did, perhaps it would be *Crescit Eundo In Numine,* meaning "It Grows as It Goes in Providence." This would be a combination of New Mexico's *Crescit Eundo* ("It Grows as It Goes") and Colorado's *Nil Sine Numine* ("Nothing Without Providence"). [237]

mountain lion. A feline without a roar; it only purrs and screams. Also called cougar, puma, panther, painter, and catamount, this is one of North America's biggest cats. Adult males usually weigh about 150 pounds; their large size and long tails distinguish them from other wild cats. As their name implies, they like mountains, but they also live in desert, chaparral, and forest habitats—any place where they can ambush other animals for a meal. Unfortunately, humans also live and play in areas where these lions prowl, and conflicts ensue. If you happen upon one, don't run away. That's what its prey do, so it may charge after you.

Instead, threaten the big cat any way you can. If you don't feel threatened, just watch, for it's a rare privilege to see one of these elusive creatures. [139]

mountain man. See **trapper.**

Mouache. A Ute band whose hunting territory spread from southeastern Colorado to Santa Fe. Their descendants are members of the Southern Ute Tribe; most live on their reservation. [119]

movies. The San Juan Basin, with its varied scenery and the Durango and Silverton Narrow Gauge Railroad, is a haven for the motion picture industry. These are some feature films with principal locations in the basin:

Across the Wide Missouri (1951, M-G-M); Clark Gable and Ricardo Montalban.

Around the World in 80 Days (1955, Michael Todd); David Niven, Cantinflas, Robert Newton.

Avalanche (1978, New World Productions); Rock Hudson, Mia Farrow.

Butch Cassidy and the Sundance Kid (1969, 20th Century-Fox); Paul Newman, Robert Redford.

City Slickers (1990, Castle Rock Entertainment); Billy Crystal, Jack Palance.

Cliffhanger (1993, Carolco/Canal+/Pioneer/RCA Video); Sylvester Stallone.

The Naked Spur (1953, M-G-M); James Stewart, Janet Leigh, Robert Ryan, Ralph Meeker.

The Outcasts of Poker Flats (1952, 20th Century-Fox); Dale Robertson, Anne Baxter.

Support Your Local Gunfighter (1971, Brigade Productions); James Garner, Suzanne Pleshette, Joan Blondell.

A Ticket to Tomahawk (1950, 20th Century Fox); Anne Baxter, Dan Dailey, Rory Calhoun, Walter Brennan.

Viva Zapata! (1952, Daryl F. Zanuck); Marlon Brando, Jean Peters, Anthony Quinn.

The basin's varied scenery and narrow-gauge railroad have provided settings for scores of movies. In this Animas Valley scene, James Stewart, Janet Leigh, Robert Ryan, and Ralph Meeker are filming *The Naked Spur*. Courtesy La Plata County Historical Society.

When the Legends Die (1972, Fox-Rank/Sagaponack); Richard Widmark, John War Eagle. [218, 256]

mule. This beast of burden was in high demand on the frontier because it is good at scrounging sparse food from the countryside. As a sterile hybrid derived from a male donkey breeding a mare, it is also more surefooted, sore-resistant, and disease-immune than a horse. It proved superior to other equine animals for packing and replaced the eastern grain-fed horse as the draft animal of choice.

Traders traveling through the basin on the Old Spanish Trail often escorted extra mules to barter for goods at the end of the trip. Initially used as pack animals, they proved to be good wagon-pullers as well. When William Becknell loaded the first wagons to head west on the Santa Fe Trail in 1821, he hitched them to horses. But once in New Mexico, he found California mules and saw their advantages. Missourians soon recognized the market for mules, and their state became a principal producer. Missouri mules brought premier prices; even poor stock often brought more than horses. The animals were employed extensively by the military as well as settlers, mail carriers, stagecoach outfits, and freighters. The army employed mules by the thousands. When Fort Lewis relocated from Pagosa Springs to the La Plata River valley south of Hermosa, the army used six-mule teams to pull its 100 wagons and brought 300 more to have in reserve—that's 900 mules! [197, 223]

museums. These are the basin's historical and cultural museums (excluding exhibits at national parks and monuments):

Anasazi Heritage Center, 27501 Colorado 184, near Dolores, administered by the Bureau of Land Management.

Animas Museum, 3065 West Second Avenue, Durango, operated by the La Plata County Historical Society.

Aztec Museum Pioneer Village, 125 North Main Avenue, Aztec.

Cortez Museum and Cultural Center, 25 North Market Street, Cortez, administered by the Cortez Center and the University of Colorado.

Farmington Historical Museum, 302 North Orchard Avenue, Farmington, managed by the Farmington Museum Foundation.

Galloping Goose Museum, 420 Railroad Avenue, Dolores, operated by the Galloping Goose Historical Society.

Mancos Museum, 171 Railroad Avenue, Mancos, operated by the Mancos Historical Society.

San Juan County Archaeological Research Center at Salmon Ruins, 6131 U.S. 64, east of Farmington, operated by the San Juan County Museum Association.

San Juan County Historical Society Museum, 1557 Greene Street, Silverton.

San Juan Historical Museum, First Street and U.S. 160, Pagosa Springs, operated by the San Juan Historical Society.

Southern Ute Cultural Center and Museum, Colorado 172 north of Ignacio, operated by the Southern Ute Tribe.

mustang. The offspring of feral horses first brought to the West by sixteenth-century Spaniards. Troops of these semiwild horses sometimes overrun the public lands, decreasing wildlife forage. They lack the stamina to be cow ponies and, if captured, are broken for less strenuous tasks or just kept as pets. Wild mustangs still roam in the basin's Disappointment Valley, northern Dolores County, and Carson National Forest. Because their burgeoning numbers threaten wildlife habitat, the Bureau of Land Management captures wild mustangs and offers them for adoption. [223]

Nageezi, San Juan County, New Mexico. A trading post with this name was started here by Jim Brimhall. The community is thirty-eight miles south of Bloomfield on New Mexico 44. [87]

Nance, Tom (ca.1860–1885). One of vigilantes of the Farmington Stockmen's Protective Association who shot down gunslinger Port Stockton at his cabin across the Animas River from present-day Flora Vista in 1881. His behavior and violence went beyond anything Hollywood dares invent. The Two-Cross Ranch cowboy was a ruffian of the worst order who flaunted his hooliganism by breaking up social gatherings. At a lantern-slide show of a passion play north of Farmington, he shot up the screen. At Parrott City he broke up dances at the courthouse by riding into the place and shooting off his revolver, forcing the townspeople to move their capers to the second floor.

When Bill Thompson stepped up to stop a cowboy row at Durango's Clipper sporting house, Nance went after him with a knife. After slashing Thompson's face, he made an enormous cut across his abdomen. Thompson slumped, his entrails spilling out onto the floor. A quick-thinking onlooker gathered up the innards in his hat as bystanders lifted Thompson up to a billiard table. A local doctor arrived to replace the intestines and stitch up the incision. Thompson lived, but we don't know whether his assailant was accused of a crime.

We do know what happened to Nance in Holbrook, Arizona. After he became an employee of the Aztec Cattle Company, he joined some of that outfit's stock buyers on a trip. While the others were out on the range choosing cattle, Nance wandered into a saloon and, as usual, got into a fight. His opponent got him down and stepped on his neck to hold him there. He then spurred Tom Nance to death. [85, 166]

Narbona (1766–1849). Navajo leader, warrior, and seeker of peace. As a youth he witnessed the slave-raiding and territorial conflicts between his people and the Spaniards. When the Spanish took the Navajo's sacred Turquoise Mountain in 1800 and set up their Cebolleta military settlement, Narbona was among the leaders who led a futile siege against them.

Navajo peace treaties with Spain, aimed at stopping slave raiding by both sides, were short-lived. After the Spanish in America established New Spain in 1821 and declared it to be the Republic of Mexico in 1824, such treaties proved to be no more durable. Dissension came not just from the Mexicans' quest for Navajo slaves. They also employed drastic measures to bring the natives into the fold of the Catholic Church. In 1823 New Mexican governor José Antonio Vizcarra refused to give up Navajo captives until all Navajos agreed to become Christians.

When, in spite of Narbona's peace efforts, slave raiding continued, he led his people's resistance. In 1835 he was the leader at a place the Navajos called Copper Pass (later named Washington Pass) when some 200 warriors emasculated a thousand-strong band of New Mexicans.

The occupation of New Mexico by the United States in 1846 did not improve the Navajos' situation; Narbona was simply forced to deal with yet another set of invaders. When General Stephen Watts Kearny sent a delegation to invite native leaders to a peace parley, Narbona used 2,000 mounted Navajos in a show of force. The treaty that followed, calling for a mutual exchange of slaves and prisoners, was never ratified by Congress.

In 1849, after riding thousands of miles to engage in dozens of peace talks, Narbona and his lieutenants agreed to peace terms set forth by James Calhoun, New Mexico Territory's superintendent of Indian affairs. At the end of the meeting, however, a dispute over an allegedly stolen horse led Colonel James Macrae Washington to order small arms and cannon fire on the Navajos. The barrage killed Narbona. Two of his sons prepared his body for burial, dropped it and Narbona's belongings into a deep crevice, then mourned four days on a nearby hill.

Narbona's murder gave credence to the beliefs of the more militant leaders, including his son-in-law, Manuelito, that peace efforts were hopeless. In the conflicts that followed, the army's scorched-earth strategy destroyed Navajo resistance. [64, 82]

Narraguinnep Reservoir State Wildlife Area. The Dolores River feeds this 535-acre body of water through an irrigation diversion system. Piñon- and juniper-covered hills surround the lake, which is situated at an elevation of 6,680 feet and keeps a residual pool to maintain fisheries.

The Colorado Division of Wildlife stocks the reservoir with walleye, yellow perch, northern pike, black crappie, bluegill, and channel catfish.

Narraguinnep Reservoir is eleven miles northwest of Cortez on U.S. 666 and two miles east on Colorado 184. [250]

Naschitti, San Juan County, New Mexico. Colonel James Macrae Washington's 1848–1849 expedition against the Navajos paused at this place, then known as Badger Springs. According to one civilian's diary, the bravest New Mexican militiaman during an encounter with the natives had painted himself and his mule entirely red.

Naschitti is forty-five miles south of Shiprock on U.S. 666. [82]

national forest system. President Benjamin Harrison blazed the start of the nation's forest system in 1891 when he set aside the Yellowstone Park Timber Land Reserve (now part of the Shoshone and Teton National Forests) under the Forest Reserve Act. Before, Congress had carelessly transferred public lands to private ownership as a means of encouraging settlement and raising revenue. President Theodore Roosevelt and his Forest Service director, Gifford Pinchot, guided the dedication of lands in the forest system that now includes 256 sites managed by the National Forest Service: 156 forests, 19 grasslands, and 71 experimental forests. The 3,000-square-mile San Juan National Forest (part of the San Juan–Rio Grande National Forest) lies within the San Juan Basin, as does the Jicarilla Ranger District of Carson National Forest. [230, 267]

national park system. While the Washburn-Langford party was exploring what is now Yellowstone National Park in 1870, a member observed that it should be preserved for the public benefit. That idea led to the establishment of Yellowstone as the nation's first national park two years later and the National Park Service in 1916. In the interim the 1906 Lacey Antiquities Act authorized the president to create national monuments by proclamation. (National parks require an act of Congress.)

Today the national system of parks, monuments, parkways, preserves, shores, and recreation areas includes 365 units. The basin holds Aztec Ruins, Hovenweep and Yucca House National Monuments, Chaco National Historic Park, and one of the system's most famous parks, Mesa Verde. [21, 230]

National Wilderness Preservation System. Under the Wilderness Act of 1964, four federal agencies administer the 148,000 square miles of public lands in the National Wilderness Preservation System: the Fish and Wildlife Service, Forest Service, Park Service, and Bureau of Land

Management. The legislation (P.L. 88-577) recognizes wilderness as "an area where the earth and its community of life are untrammeled by man, where man himself is a visitor who does not remain."

The basin has five wilderness areas: Bisti, De-Na-Zin, Lizard Head, South San Juan, and Weminuche.

natural gas. The extraction of this energy source from the coal seams beneath the basin, and the economic and social consequences of the industrialization that followed, paralleled that of the oil industry. An 1896 well, the first in New Mexico, produced gas, but the first commercial use came at Aztec in 1921. A pipeline to California, completed in 1951, opened major markets and helped jump-start the 1950s energy boom. [55]

Navajo Indian Irrigation Project. By treaty, the United States established the Navajo Reservation in 1876 and promised to provide farmland for the Navajo Tribe. Since most of the reservation receives less than nine inches of rainfall yearly, its lands cannot be farmed successfully without irrigation water. This Bureau of Reclamation project, started in the 1960s, provides that water. Located on an elevated plain south of the San Juan River near Farmington, this 110,630-acre system receives water from Navajo Reservoir some thirty miles east. In 1976 water began flowing from the lake through a pipe seventeen feet in diameter—the largest and heaviest precast concrete pipe made. The Navajo Tribe operates and maintains the project. [154, 262, 265]

Navajo Lake State Park (New Mexico). Both the Pine River and San Juan River portions of this facility, operated by the New Mexico State Parks and Recreation Division, stretch along the banks of Navajo Reservoir. To get there from Aztec, follow New Mexico 173 and New Mexico 511 twenty-one miles east; from Bloomfield, go twenty-three miles northeast via U.S. 64 and New Mexico 511. [227]

Navajo Lake State Recreation Area (Colorado). On the Piedra River branch of the Navajo Reservoir, this facility is operated by the Colorado Division of Parks and Recreation. The area is thirty-seven miles southwest of Pagosa Springs via U.S. 16 and Colorado 151.

Navajo Mine. Like the La Plata and San Juan Mines, this mine feeds coal to the basin's giant power plants. From it, behemoth vehicles traverse a haul road to the Four Corners Generating Station. The mine is headquartered three miles south of Waterflow.

Navajo Nation. The Navajo Tribe and its reservation.

Navajo Reservation. After the United States recognized the cruel fallacy of imprisoning the Navajo people at Bosque Redondo, Congress permitted them to return to a portion of their homeland. That portion is now their reservation—the largest in the country. Established in 1867, it encloses some 25,000 square miles, a region the size of West Virginia, stretching through New Mexico, Arizona, and Utah. About a fifth of this rugged, semiarid land lies within the basin. The reserve has a population of 167,000 (1990 census). [82, 220]

Navajo Reservoir. One of the four storage units of the Colorado River Storage Project. (The other three are Glen Canyon in Arizona, Flaming Gorge in Utah, and Wayne N. Aspinall near Ridgeway, Colorado.)

The lake's fingers extend thirty-five miles up the San Juan River, thirteen miles up the Pine River, and four miles up the Piedra River. Under its 15,610-acre surface, the reservoir holds more than 1.7 million acre-feet of water for release to the Navajo Indian Irrigation Project and other agricultural, municipal, and industrial users. The Bureau of Reclamation built the dam between 1958 and 1962; the Bureau of Land Management is responsible for the Simon Canyon Recreation Area below the dam, a premier trout-fishing spot. [41, 154, 257, 265]

See also **Navajo Lake State Park; Navajo Lake State Recreation Area;** and **Navajo State Wildlife Area.**

Navajo Reservoir Management Area. This 218,200-acre tract is the subject of what the bureaucrats call a National Performance Review Laboratory—an experimental arrangement in which all agencies with land management, wildlife, or recreation functions cooperate to manage the area's natural resources. The following list of agencies shows the complexity of managing western public lands. Participating are federal agencies—Bureau of Land Management, Bureau of Reclamation, Fish and Wildlife Service, Geological Survey; Colorado agencies—Oil and Gas Conservation Commission, Division of Wildlife, Division of Parks and Recreation, Division of Water Resources; New Mexico agencies—Department of Game and Fish, Parks and Recreation Division; and tribes—Navajo, Jicarilla Apache, Southern Ute. [266]

Navajo rug. No craft reflects more the influence of other cultures on the Navajos or better displays their love of beauty than the Navajo rug.

The use of wool to make clothing started in the Southwest among the Pueblo natives after the Spanish introduced sheep to the Rio Grande valley during the sixteenth century. Some Puebloans who fled from the

Navajos, ca. 1910. These Native Americans were descendants of natives forced by Colonel Kit Carson to take the "Long Walk" to Bosque Redondo. After the confinement plan failed, Congress granted the Navajos a reservation. Courtesy Fort Lewis College, Center of Southwest Studies.

Spaniards after the Pueblo Revolt of 1680 took refuge among the Navajos. From the Pueblo natives the Navajos learned the art of weaving. As Navajo slaves made blankets for their Spanish masters during the next century, they used both Navajo and Spanish techniques. By the 1860s the Navajos were trading for colorful European and American yarns to enhance their designs. The federal government's confinement of the Navajos at Bosque Redondo during that decade deprived them of their traditional wools and forced them to use the machine-made yarn provided by their captors. After the Navajos were permitted to leave Bosque Redondo and as the twentieth century approached, weavers found a market for heavier fabrics, rugs for the whites. With the evolution to rugs, regional styles of weaving matured. Today the Navajos' commercial fabrics include general rugs, regional style rugs, saddle blankets, and specialty rugs.

By adopting the Pueblo loom, using the wool of Spanish and French sheep, and integrating European yarns into their tapestries, Navajo weavers have used features of other cultures to form highly prized works of art. [71]

Navajo State Wildlife Area. A 520-acre tract of bottomland on the northwest edge of Colorado's Navajo State Recreation Area at Navajo Reservoir. At 6,100 feet elevation, the area has over four miles of streams along the tributaries of Sambrito Creek, amid western wheatgrass, ricegrass, rabbit brush, grama grass, and sagebrush with willows and cottonwoods along the drainages. The Colorado Division of Wildlife leases the parcel from the Bureau of Reclamation for waterfowl and small game production. The area is good for prairie-dog hunting. Some warmwater fishing is available in the reservoir's Sambrito Bay.

The area is four miles west of Arboles on Colorado 151 and a mile south. [250]

Navajo Trail. A name sometimes used for U.S. 160, a route from Chicago to Los Angeles. It cuts through the Colorado portion of the basin.

Navajo Tribe. After migrating from what is now western Canada in the middle of the eleventh century, the Athapascan people who formed this clan settled on the land surrounded by the Rio Grande, San Juan, and Colorado Rivers.

Starting in the sixteenth century, Navajos and their traditional enemies, the Apaches, engaged in a deadly relationship with invading Spaniards. The conflicts continued after Mexico's 1821 revolution. The natives raided settlers in Old Mexico as well as New Mexico, stealing sheep, cattle, and horses and taking captives to meld into their tribe. In return, the Spaniards and Mexicans plundered the Navajos and impressed some of them into slavery. Mexican traders played both sides and made money from the conflict by providing arms and ammunition. This pattern was well established before Anglo-Americans moved west into the region and before the United States officially entered the picture by declaring war on Mexico in 1846.

The Hispanic and Navajo raiding and plundering of each others' people set the foundation for the United States' war with the Navajos. For as the Navajos had raided and pillaged the Spanish and Mexicans, so they did the intruding Anglos.

When Colonel Stephen Watts Kearny took over the territory from Mexico in 1846, he promised the Hispanics protection from hostile Native Americans—he offered no reciprocal assurance to the Navajos. To carry out Kearny's pledge, the army built forts at key locations along rivers and trade routes. Territorial governors, who also served as superintendents of Indian affairs, distributed rations to the Navajos and encouraged farming. Thus, the tribe and the army managed to reach a truce in 1861. But it was short-lived. While the two sides were having a horse race that year, the Navajos accused the army competitors of cheating and stealing a horse. In the following melee soldiers opened up with howitzers and killed twelve Navajos. In response, the natives resumed their guerilla war.

As the Civil War raged in the East, General James H. Carleton and his Californians arrived to hold New Mexico for the Union. In 1863, with New Mexico free of Confederates, he turned his army on the marauding natives. To subdue them, he chose his old friend Kit Carson, now colonel of the First New Mexico Cavalry. After Carson subdued the Mescalero Apaches and sent them to Bosque Redondo, Carleton turned his attention to the Navajos. He concluded that they too should be removed and become farmers at the internment. He gave the Navajos an ultimatum: report to Bosque Redondo. To enforce the order, he again called on Kit Carson.

Subjugating the Navajos was no small task. Unlike the 500 mountain-dwelling Mescaleros Carson had subdued, the 12,000 Navajos ranged from the Continental Divide to the Colorado River. To break the Navajos, Carleton and Carson advanced a scorched-earth policy. From the San Juan Basin south to the Little Colorado, Carson marched his troops in successive sweeps, destroying everything useful along the way—field crops, orchards, hogans, livestock. He then sacked the Navajos' supposedly impregnable and sacred fortress, Canyon de Chelly.

As thousands of Navajos surrendered, Carson's troops escorted them to Bosque Redondo, in some cases a 300-mile forced march. Along the way, hundreds died. This was the Navajos' "long walk." The 8,000

Navajos entrapped at Bosque Redondo in the Pecos Valley found few provisions, infertile soil, and disease. Free-roaming Comanches stole their livestock. The 4,000 who fled west under Manuelito fared little better than those who surrendered. In 1866 they also gave up.

Two years later, several Navajo chiefs were permitted to plead their case in Washington, D.C., where they persuaded officials to grant them a reservation. After four years of confinement, the Navajos were allowed to occupy again a portion of their old homeland. That portion is now their reservation.

They had been punished for making war on Spaniards and Mexicans, or so they were told. Kearny quelled the natives to gain Hispanic support. A possible ulterior motive worsens the baseness of Carleton's treatment of the Navajos, for he saw "fields of gold and other precious metals" resting under their homeland—riches to be gathered when they were removed.

As the Navajos made the reverse march back to their homeland, they promised to remain at peace with their Anglo-American conquerors if they would have teachers for their children and token replacements for their slaughtered livestock.

Today perhaps a fifth of the 167,000 Navajos live within the basin's portion of their rugged, semiarid reservation. [82, 123, 245]

Needle Mountains. The Animas River valley cuts through this southwestern buttress of the San Juans, centered some thirteen miles south of Silverton in the Weminuche Wilderness. Within a twenty-five-square-mile section, the range has ten peaks over 13,000 feet high. [244]

nester. A settler who followed the cattlemen into the West with the intent of establishing a farm on public lands, often through homesteading. The nesters' migration gave rise to the conflicts between cattlemen and homesteaders—the plot for numerous Western novels and motion pictures. [223]

Newcomb, San Juan County, New Mexico. This Navajo trading post, like many others, got started after the Navajos were permitted to return from captivity at Bosque Redondo near Fort Sumner. Started in 1868, it was the Blue Moon Trading Post before Arthur J. Newcomb bought it in 1913. Also called Nava, it was once in a chain owned by Gallup trading-post magnate C. C. Manning. Under Newcomb's management, and through his wife's friendship with medicine man and rug weaver Hosteen Clah, this became one of the most successful posts in the Chuska Valley.

Newcomb is thirty-one miles south of Shiprock on U.S. 666. [71, 87, 93]

New Mexico. Sixty percent of the basin's 150,000 people live in this state, the Land of Enchantment, which joined the Union in 1912. Like Florida, it is also called the Sunshine State. The basin's New Mexico residents make up 6 percent of the state population of 1,515,069 (1990 census); New Mexico land within the basin comprises a similar portion of the state's whole.

The state's symbols: flower, yucca; tree, piñon; animal, black bear; bird, roadrunner; fish, cutthroat trout; vegetables, chili and frijol; gem, turquoise; colors, red and yellow; song, "O Fair New Mexico"; Spanish-language song, "Así es Nuevo Mejico"; poem, "A Nuevo Mexico"; grass, blue grama; fossil, *Coelophysis;* cookie, bizcochito; insect, tarantula hawk wasp. [237]

New Mexico Territory. Two years after Mexico ceded a vast area of the American Southwest to the United States at the conclusion of the Mexican-American War in 1848, Congress designated almost half the region as New Mexico Territory. It stretched from Texas to California south of the 37th Parallel (the same line became the Utah-Arizona and Colorado–New Mexico borders). The territory grew south when the Unites States got more land from Mexico with the Gadsden Purchase of 1853. Distrusting eastern Mew Mexico's Confederate sympathies, Congress split off the Union-secure western part of the territory in 1862 to form Arizona Territory. These acts set the boundaries for what would become the forty-seventh state in 1912. [78, 204]

oil. An energy source that has prompted economic booms throughout the San Juan Basin ever since the Hogback Oil Pool was discovered west of Farmington in 1922. In the early 1950s, oil production firms swarmed to Durango, creating the city's first postwar population spurt, but a shortage of sizable discoveries and a lack of oil pipelines brought the boom to a whimpering halt. Elsewhere in the basin, however, the explosion continued. Energy companies pumped oil from pools in New Mexico's portion of the basin through a new pipeline to California. Farmington exploded from an agrarian town into an industrial city; during the decade of the 1950s the community experienced a fivefold increase in population. Although changes in federal policies dampened the oil industry in the 1960s, petroleum-related prosperity continued, swelled by the so-called oil crisis of the 1970s. Farmington's economy took a dive in the mid-1980s, however, when the price of oil plunged from thirty-six dollars a barrel to ten. Nevertheless, oil exploration, production, and transportation remain an essential part of the basin's economy. [55, 204, 215]

Old Spanish Trail. Some trails are more famous, but none was more arduous than this route that cut through the San Juan Basin on its way from Santa Fe to Los Angeles. Sometimes called merely the "Spanish Trail," it was so long and rugged that wagon transport was out of the question; goods moved along its 1,200 miles on pack animals.

From what we know, no single explorer blazed its course; rather it was advanced by a succession of travelers. Juan Marie de Rivera used its eastern section when he journeyed out of Santa Fe in 1765 to explore the Gunnison River country north of the basin. Domínguez and Escalante followed it as they sought a route to California in 1776. In the winter of 1829–1830 Mexican trader Antonio Armijo made the first

Oil drilling rigs like this one near Aztec also sprouted around Farmington and Cortez in the 1920s, but the basin's biggest energy boom came three decades later. Courtesy La Plata County Historical Society.

known journey of the whole trace. A year later William Wolfskill initiated an era of commerce by proving that pack trains could traverse the trail.

For the next dozen years New Mexicans used the trail as a trading route to bring goods, some of Chinese origin, and mules from California. Some of the goods moved on to the Missouri market by way of the Santa Fe Trail to Independence. In this sense, each trail extended the other to form a 2,000-mile route from east of the Missouri River to the Pacific Ocean.

Like other trails through uncharted wilderness, the route likely followed a number of tracks as freighters sought easier paths. But in this case no wagons cut ruts to ease the task of historians, so we must rely on the accounts of Spanish explorers. The Domínguez and Escalante Expedition represents the trail's course through the basin. The party headed north out of Santa Fe and followed the Rio Grande and its western affluent, the Rio Chama, past Abiquiu before turning north to avoid the stream's rugged canyon. After rejoining the river, they proceeded past Tierra Amarilla, then northwest over the Continental Divide to enter the basin near modern-day Monero.

Once in the basin, their track went past the sites where you now find Lumberton and Dulce, then to the San Juan River some five miles below its confluence with the Navajo. They followed the San Juan to the mouth of the Piedra (now submerged by Navajo Reservoir) and traversed a route northwest past present-day Ignacio, a course the Denver and Rio Grande Railway would take a century later. From there the expedition continued northwest across the Pine, Florida, and Animas Rivers, crossing the last a few miles south of present-day Durango. The party moved west through Ridges Basin (a swale west of the Animas River marked for flooding by the Animas–La Plata Project), passed over Hesperus Pass and the future sites of Hesperus, Mancos, and Dolores, then followed a course near present-day Colorado 184 (Domínguez and Escalante Memorial Highway) to leave the basin north of Dove Creek.

The trail went on to California by swerving north to avoid the Grand Canyon, then sweeping southwest to Los Angeles. The trace was more than a route for commerce. During the 1846–1848 war with Mexico, it served as a communication route so the United States could keep track of goings-on in California. Military courier Kit Carson twice carried dispatches over its entire length. News of gold strikes carried on his second trip east set off a rush of wealth seekers in 1849. They became known as the Forty-Niners.

After the United States wrested from Mexico the vast territory linked by the trail, other routes beckoned the trader. The nation's westward expansion soon moved along easier routes and federal wagon roads. Lack of use and time have obliterated most of the Old Spanish Trail. [57]

Oñate, Juan de (ca. 1550–ca. 1624). Spaniard whose expedition up the lower segment (Rio Abajo) of the Rio Grande valley established the initial Nuevo Mexico in 1598. [19, 246]]

108th Meridian. A north-south orientation line running east of Durango and through Aztec and Bloomfield. The 100th Meridian, which runs through Dodge City, Kansas, over 500 miles to the east marks the beginning of the arid West, where rainfall averages sixteen to twenty-eight inches per year. Some places, among them the lower San Juan Basin, receive even less. [117]

Ophir Pass. The Utes traveled from the San Miguel Valley to the Animas Valley through this swale that separates the San Miguel River and Mineral Creek drainages. It has since been used by trappers, miners, freighters, and hikers. It was the scene of a tollroad in 1881, and when Otto Mears built the Ophir Loop of the Rio Grande Southern Railroad near the pass in 1891, the section became known as "Mears's Puzzle," a marvel of engineering and persistence. North of the pass, just outside the basin, lies the town of Ophir. Both place-names came from the biblical location of King Solomon's mines. The road over the pass provides a summer connection between Silverton and Telluride.

Ophir Pass (elev. 11,789 feet) is accessed via Forest Service 679, which leaves U.S. 550 five miles north of Silverton. [175, 219]

Ouray (ca. 1833–1880). The Ute chief who relied on diplomacy, more than warfare, to keep the whites at bay.

The son of a Ute-adopted Jicarilla Apache, Ouray grew up working as a sheepherder for a well-to-do Spanish family near Taos, New Mexico. Whether his family's design was for young Ouray to provide income or become educated in the white man's way, we don't know. In any case, by the time he rejoined his father among the Tabeguache (later called the Uncompahgre) Utes, he had learned four languages. Through this ability, combined with his dignity and charm, he became known as an unusually "civilized" Native American. He became chief of the Uncompahgre band, and even though centralized power was foreign to the tribe, as it was to most native tribes, Ouray exercised enough control to become the Ute's main spokesman. U.S. negotiators regarded him as chief of all the Utes.

Recognizing the futility of waging war against the overwhelming power of the invading hordes of Europeans, Ouray urged restraint. By strength of personality, tact, and diplomacy, he shrewdly negotiated agreements that were more beneficial than those that might have come through warfare. While the United States was herding most tribes to reservations, he helped negotiate the Treaty of 1868, which promised to keep the Anglos out of the Rockies' West Slope. Though he let whites come in for mining purposes by consenting to the San Juan Cession, or Brunot Agreement, in 1873, the Treaty of 1868 stretched his people's freedom another twelve precious years.

An occurrence at the White River Agency in 1879 provided the United States an excuse to drive the Utes off their West Slope hunting grounds. Commonly called the "Meeker Massacre," it destroyed his lifelong efforts to maintain peace while keeping a sizable territory for the Ute bands. The next year, in his teepee near Ignacio, Ouray died despondent. For twenty years he had matched wits with generals, Department of the Interior secretaries, bureau chiefs, governors, presidents, and Congress to postpone the day when his shy people would be driven from their awe-inspiring lands. [86, 119, 246]

Out West. One of the four epic American directions. As "Back East" conjures a crowded city, "Down South" an antebellum plantation, and "Up North" a snowbound farm, so does "Out West" bring a vision of wide-open spaces bordered by glorious mountains dotted with wandering trappers, raucous cowboys, yelling Indians, and blue-coated cavalrymen riding to the rescue—features all endemic to the San Juan Basin. No map prescribes these quadrants, but Out West usually means the territory west of the Mississippi River, especially that section beyond the 100th Meridian. During the nineteenth and early twentieth centuries no phrase did more to stir the dreams of a city dweller. We should not be surprised if it stirs the imagination of urbanites through the twenty-first century as well. [47]

Oxford, La Plata County, Colorado. This farming community got started in the late nineteenth century on the route of the Denver and Rio Grand Railway and saw an influx of settlers after the Ute Strip opened for white settlement in 1899. Oxford is eighteen miles southeast of Durango via U.S. 160 and Colorado 172.

pack rat. See **woodrat.**

Pagosa Junction, Archuleta County, Colorado. Before lumber companies built the Rio Grande, Pagosa and Northern Railroad to branch off the Denver and Rio Grande Railway at this place in 1900, it was called Gato. The lumber railroad went from here to Pagosa Springs. After the forests were stripped and the railroads were pulled up, the community faded.

Pagosa Junction is twenty-five miles southwest of Pagosa Springs via U.S. 160 and Archuleta 500. [91]

Pagosa Springs. *Pagosa* is Ute for "healing waters" and here is the Ute legend about these hot springs.

A plague fell upon their tribe, and their medicine man could not quell it. All of their wild plant and animal medicines failed to curb the malady. Death came to many members of their dwindling tribe. On the banks of the San Juan River their leaders called a great council and built a huge fire to send a message to their god, the Chief of the Happy Hunting Ground. They danced, told of their braves' exploits, admitted their sins, and implored the Great Chief to heal them.

While the Utes slept that night, at the spot where the fire had died to embers, a pool of boiling water appeared. They bathed and drank the water, cried their chants of victory, and thanked the Great Chief for answering their call. Long did the Utes camp at the healing pool and whoop, "Pagosa! Pagosa!"

There is an epic story about how the Utes and Navajos decided which tribe should possess the healing springs. The tribes fought bitterly for the site of the Pagosa until 1867, when they agreed to settle the issue with a duel between a chosen brave from each tribe. So well did the Utes like Indian agent Colonel Pfeiffer of the New Mexico Militia that they admitted him to their tribe. The colonel offered to fight as a Ute if the duel were with Bowie knives. The Navajos agreed.

In a scene that could have been written for Hollywood, the duelers stripped to the waist and circled each other, one arm tied behind—the small, wiry colonel and the brawny Navajo. Quickly the colonel maneuvered to a position of advantage and thrust the heavy knife into the heart of his stalwart opponent. The Navajos acknowledged defeat; the Utes took possession of the healing waters. But for the Utes, who measure time by generations, the victory had a short life. Within a couple of decades, their healing waters were taken over by white settlers.

The springs are at the city of the same name. [185]

Pagosa Springs, Archuleta County, Colorado. Although a tollroad leading to the Animas Valley passed Pagosa Springs as early as the 1860s, permanent white settlement didn't get started until 1878 when the army began construction of Fort Lewis. Within a year the town boosters were looking forward to the arrival of the Denver and Rio Grande Railway. The army post, an Indian agency, and the hot springs seemed to assure a rosy economic future.

But the railroad passed twenty-seven miles south. Not only did it miss Pagosa Springs, by replacing animal transport it reduced the number of freight wagons that clattered through town. Then Fort Lewis moved west and the Indian agency left as well. To compound their difficulties, the residents were confronted with land ownership questions—but the problems were eventually resolved amid the usual frontier confusion of political manipulation and speculator skulduggery.

By the 1890s, timbering operations were in full swing as branch railroads south of the town carried lumber to the D&RG's main line or east over the Continental Divide to Chama. As the forests were cleared, the prosperity of lumbering faded. But the economy got a boost when Wolf Creek Pass opened for automobile traffic in 1916.

Pagosa Springs (1990 pop. 1,207; elev. 7,079 feet) is fifty-nine miles east of Durango on U.S. 160. [91]

Palmer, William Jackson (1836–1909). The Quaker who brought the Denver and Rio Grande Railway to the San Juan Basin.

Reared in Delaware and trained as an engineer, Palmer interrupted his railroad career as the confidential secretary to the Pennsylvania Railroad to fight for the Union in the Civil War, advancing to the rank of brevet brigadier general. Thus, he is often called "General" Palmer.

To become an army officer, he formed a cavalry unit, then got it commissioned by Pennsylvania and accepted into the Union army. Palmer fought at Shiloh, led the Fifteenth Pennsylvania Cavalry against the Confederacy's invasion of Pennsylvania, helped the Federals win at Antietam, and commanded a regiment at Chichamauga.

After the war he became manager of surveys for the Union Pacific Railroad, Eastern Division (later called the Kansas Pacific) as it laid rails from Kansas City into Nebraska. His personal survey of the West convinced him that the railroad could build a line through New Mexico to San Francisco. (The Kansas Pacific never built such a road, but three other companies extended their lines as Palmer had proposed.)

On his own, he borrowed money from eastern investors and started the Denver and Rio Grande Railway to run a route from Denver to Mexico City and another from Denver to Ogden. In 1879 he started a branch from the Denver–El Paso line at Alamosa, on Colorado's East Slope, and headed for the San Juan Basin.

Palmer did more than build American railroads. Even before he organized the D&RG, his Colorado Land Company had started buying property near Pike's Peak that would become the site for Colorado Springs. He followed the pattern with the Durango Land and Coal Company, which bought land and platted Durango. Another of his products was the Colorado Fuel and Iron Company in Pueblo. Not content with building in the American West, Palmer organized railroads in Mexico. Officials of that country placed a bronze tablet in his honor at the Colonia Station in Mexico City.

Palmer spent his last years crippled by a horseback-riding accident before he died at age eighty-three.

William Jackson Palmer does not deserve the image of a railroad baron. When he got an unexpected $1 million from the sale of his interest in the D&RG, he quietly distributed it among the line's workers— without regard to rank. He bought 753 acres, now called Palmer Lake, as a gift to Colorado Springs and endowed Colorado College and the Hampton Institute of Virginia. Durango's public library at East Second Avenue and Twelfth Street rests on property Palmer donated for that purpose. [22, 46]

Parrott City, La Plata County, Colorado. In 1873 John Moss traded one hundred ponies and a load of blankets for Ute chief Ignacio's permission to use thirty-six square miles at the mouth of La Plata Canyon for placer mining. There he promoted a town and named it to honor the venture's financiers, members of the Tiburcio Parrott family of San Francisco.

Although most worthwhile strikes came farther up the valley, in the 1870s the town had its heyday. Migrant scouts, prospectors, freighters, trappers, and cowboys mixed with its 500 residents to produce a robust community. Beginning with the formation of La Plata County in 1876, the town served as the county's seat of government. But Moss's mining ventures failed to pan out, and the basin's maiden railroad came to a different part of the county. After the voters moved the county seat to Durango in 1881, Parrott City faded away.

The obliterated site of Parrott City is north of U.S. 160 on La Plata 124 near Hesperus. [48, 49, 208]

Pastorius Reservoir State Wildlife Area. This forty-nine-acre irrigation impoundment, operated by the Colorado Division of Wildlife, lies southeast of Durango in an agricultural setting at 6,860 feet elevation. Cottonwoods, alders, and willows border the shallow lake that is home to rainbow trout, northern pike, bluegill, yellow perch, channel catfish, and largemouth bass. Waterfowl and shore birds use the reservoir for spring migration. From Durango, Pastorius Reservoir is eight miles east on U.S. 160, two miles south on Colorado 172, a mile south on La Plata 302, and a half mile west on La Plata 304. [250]

Pendleton. See **La Plata, San Juan County, New Mexico.**

Perins Peak. A peak overlooking Durango to the west named after a surveyor who laid out the Durango townsite for the Durango Trust, an offshoot of the Denver and Rio Grande Railway. [113]

Perins Peak State Wildlife Area. Lying northwest of Durango, this area consists of several Bureau of Land Management and Colorado Division of Wildlife tracts totaling 6,900 acres at elevations between 6,800 and 8,700 feet. Its terrain is hilly; cliffs surround its several drainages. Gambel oaks and ponderosa pines shelter meadows of native grasses while aspen and Douglas firs dominate its north slopes. Deer, elk, and bears use the area, as do band-tailed pigeons, doves, rabbits, blue grouse, and peregrine falcons.

From Durango, the area is four miles west on U.S. 160, then north on La Plata 207 and La Plata 208 (or five miles west on U.S. 160, then north through private land). [250]

peyote. A cactus used by some Navajos and other Native Americans for ceremonial purposes and for its mind-altering qualities. The plant's aboveground portion is gray-green and looks like a bunch of carrot tops. Peeled and chewed, peyote roots produce pleasant visions and trances. The potent root also makes a healing tea. [223]

Piedra River. This stream's bed is a rocky hiking trail during low water, and its banks are a series of rock walls. It flows sixty miles from within the Weminuche Wilderness to join the San Juan where the latter has been stilled by Navajo Reservoir on the Colorado–New Mexico border. The box canyons of its upper reaches challenge the most experienced river runners. There also the ruggedness of the terrain has discouraged timbering and other disturbances. Its corridor—like the river itself—invites only the hardy. [104]

Pine River. The name comes from an 1877 Hayden atlas that adopted the Spanish phrase Rio de los Piños, "River of the Pines." Some references call it simply "Los Piños." Midway along its seventy-mile course from near Rio Grande Pyramid to Navajo Reservoir at the Colorado–New Mexico border, a dam blocks it to form Vallecito Reservoir. The upper part of this river has the privilege of being remote from a highway. For twenty-five miles, as it flows within the Weminuche Wilderness, its isolation as well as its mountain features make it a truly wild river. [104]

Pine River Project. The latest in a series of irrigation schemes along the Pine River dating back to 1877. Pioneers dug irrigation ditches to deliver water for Indian agencies and farms. When the courts decreed, in 1930, that much of the water used by the settlers belonged to the Southern Utes, the Anglo-Americans were left potentially dry. The situation impelled a plan to store spring floodwaters.

The project consists of Vallecito Dam (its name is Spanish for "little valley") and its reservoir, built by the Bureau of Reclamation between 1938 and 1941. The mile-wide lake, which extends three miles up the Pine River and its tributaries, impounds 129,700 acre-feet of water to irrigate 54,000 acres stretching twenty-eight miles south—past the New Mexico border. A quarter of the irrigated land belongs to Native Americans.

The Pine River Irrigation District manages the project and, with the Forest Service, provides recreation facilities. The lake is eighteen miles from Durango by way of La Plata 240 and 501; from Bayfield, it is eleven miles north on La Plata 501. [264]

piñon. The state tree of New Mexico spreads across the basin's mesas and foothills below 8,500 feet elevation, usually mixed with juniper. Bracketed in the vegetation zone between sagebrush and Douglas fir, this nut pine is also called the New Mexico piñon pine or Colorado piñon pine. Its seeds are relished by connoisseurs and harvested by both people and animals. The number of animals that eat its seeds would make a long list; squirrels and chipmunks cache them as do various jays.

These sparse piñon-juniper (P-J) forests prevail in the basin's Upper Sonoran Zone, where the scarcity of rainfall prevents the growth of denser vegetation.

Before cattle grazed their meadows and lumbermen cut their trees, ponderosa pine forests sheltered rich natural diversity across much of the San Juan Basin's Montane vegetation zone. Courtesy San Juan National Forest.

Pine nutting has long been an important fall activity for Hispanics as well as Native Americans. The nuts enhance many international cuisines. With its resin the Apaches and the Navajos made their baskets watertight. Because it is both useful and prevalent, the piñon enjoys a role in Native American mythology.

A piñon with a trunk six inches in diameter has seen a century of seasons and, if allowed to thrive, may see a century more. Well separated from their neighbors, these trees can stand thirty feet high, but riding horseback in a piñon forest, you can see over most of their rounded crowns. [79, 233]

pinto. This Spanish word for "painted" refers to horses that are bicolored; one of the colors is white, the other brown, tan, or black. Many Native Americans as well as others prize such animals for their appearance.

The term also describes multicolored beans, such as those grown around Rico, "Pinto Bean Capital of the World." [223]

Pleasant View, Montezuma County, Colorado. A farming community started by homesteaders around 1913. Pleasant View is twenty miles north of Cortez on U.S. 666. [50]

ponderosa pine. Before the pioneers cut these great trees to build mines, railroads, and towns, they sheltered broad meadows of diverse vegetation throughout much of the West. Cutting and overgrazing reduced the ground cover, thereby promoting a thick second growth of pines that discourages diversity. Where there once might have been fifty tall pines per acre, there are now sometimes thousands that serve as fuel to promote intense fires. Without native grasses, forest floors are carpeted with pine-needle kindling. By suppressing forest fires, the Forest Service and other public land stewards have unwittingly permitted fuel to accumulate. As a result, many new-growth ponderosa pine forests are a holocaust waiting to happen. The Forest Service is exploring harvesting methods that reduce the intensity of fires by thinning small second-growth trees. This plan may restore some ponderosa pine forests to their previous grandeur.

Also called the western yellow pine, this evergreen is bisexual and cone-bearing with needles that form in bunches. In their ideal climate these evergreens can reach heights of one hundred feet and have trunks forty-five inches in diameter, but in the basin they seldom get that big. Dominant in forests from northern Mexico to southern Canada, the pines grow in the basin most abundantly at elevations of 7,000 to 9,000 feet. [200, 233]

populations. Basin populations (1990 census) of San Juan Basin counties and cities:

Counties with Territory in the Basin

Archuleta, Colorado	5,345
Dolores, Colorado	1,504
Hinsdale, Colorado (est.)	0
Mineral, Colorado (est.)	0
Montezuma	18,672
La Plata, Colorado	32,284
Rio Arriba, New Mexico (est.)	3,000
Sandoval, New Mexico (est.)	200
San Juan, Colorado	745
San Juan, New Mexico	91,605
Total San Juan Basin (est.)	**150,155**

Cities and Towns Enumerated by Census Bureau

Aztec, New Mexico	5,479
Bayfield, Colorado	1,090
Bloomfield, New Mexico	5,214
Cortez, Colorado	7,284
Dolores, Colorado	866
Dove Creek, Colorado	643
Dulce, New Mexico	2,438
Durango, Colorado	12,430
Farmington, New Mexico	33,997
Flora Vista, New Mexico	1,021
Fruitland, New Mexico	700

Ignacio, Colorado	720
Kirtland, New Mexico	3,552
Mancos, Colorado	842
Pagosa Springs, Colorado	1,207
Shiprock, New Mexico	7,687
Silverton, Colorado	716
Towaoc, Colorado	700
Waterflow, New Mexico	300

prairie dog. These rodents are a lot like people—so a scientist who studies them has concluded. They care for each other, greet with a "kiss," set boundaries, and live in towns. When they sense danger, they warn other clan members.

When ranchers first settled the West, they considered prairie dogs pests. That's because the little rascals like heavily grassed areas, as do cattle. Due to poisoning programs by ranchers and government agencies, prairie dogs now occupy only about 2 percent of the range they enjoyed at the beginning of the century. The black-tailed species predominate east of the Rockies. The basin is home to the white-tailed variety.

Although their holes can break a horse's leg and their burrowing can ruin a crop, some studies indicate that when prairie dogs disturb an area, they encourage new grass that is good for bison and cattle. To the wildlife watcher, prairie dogs display a cute amicability. But their above-ground behavior shows a different side. After the female mates with her local male and gives birth five weeks later, an epidemic of infanticide begins. Females cannibalize and wipe out each other's litters with abandon. Generally, only two out of five litters survive. [183]

prostitution. The sordid activities in the red-light districts of the basin's pioneer years were tolerated as long as they stayed in their own part of town. The city fathers of Durango put up with gambling and whores because they produced revenue for the local government and circulated money in the community. Silverton indulged its notorious Blair Street. Just as cow thievery became "rustling," and therefore somehow less serious than "stealing," so did prostitutes become "soiled doves" to avoid offense to Victorian sensibilities. Even in the 1990s, sex-for-sale's history is treated lightly by some. An issue before the Durango City Council in

The basin is home to white-tailed prairie dogs. They and their black-tailed cousins east of the Rockies now occupy only 2 percent of the terrain they enjoyed before white settlement. Courtesy San Juan National Forest.

1996 was whether to name a street "Red Light Lane" in memory of its nineteenth-century tenderloin experience.

But the myth of a cute bawdy house with a tinkling piano, affectionate girls, and a caring madam serving friendly customers is easily dispelled. Prostitution was no more glamorous on the frontier than it is today. One historian has observed of pioneer prostitutes: "Most faced lives of poverty, alcoholism, disease, violence and drug addiction. Many ended their lives by suicide." [29, 73, 110, 130]

Pueblitos of Dinetah. While most basin visitors view pre-Columbian sites by joining the crowds at national parks and monuments, some adventure to more remote places, such as the Navajo Pueblitos. Forty-eight of its seventy-six identified sites are listed in the National Register of Historic Places. They are on land managed by the Bureau of Land Management.

Puebloans and Navajos apparently lived together in these small pueblos after the Puebloans fled there from Spaniards. The Pueblo Revolt of

1680 broke the Spanish hold on the Rio Arriba (northern Rio Grande) country and started a period of social turmoil and hostility. When the Spanish regained control in 1692, some Pueblo refugees escaped to Navajo territory. That is why pottery, rock writings, and construction methods at these sites reflect Pueblo culture. Artifacts found at the site indicate the occupants had contact with natives over a wide area.

By 1715 the Utes were threatening the survival of these Pueblo-Navajo natives. For protection, the defenders built on mesa tops and cliff faces. The remnants perch on high mesas at 5,800 to 6,500 feet elevation overlooking deep, narrow ravines. With strategic views of the surrounding territory and well within sight of neighboring pueblitos, the occupants enjoyed a commanding advantage over their attackers.

The Pueblitos of Dinetah are east of Blanco two miles on U.S. 64, then south twenty miles on San Juan 4550 and San Juan 379 (Canyon Largo Road). [153]

Pueblo Revolt. Although this 1680 Pueblo victory centered along the upper Rio Grande, removed from the San Juan Basin, it is notable because it delayed Spanish conquest of New Mexico, deferred Hispanic intrusion of the basin, and changed relations between tribes and races.

For decades the natives had resented the Spaniards' imposition of their Christian religion. In their kivas they defied the Europeans by secretly practicing the forbidden rituals of their native lore, fermenting open defiance. As the Spaniards aggravated the Puebloans' hardships by taxing them to fight the Navajos, a leader named Pope came forth to incite rebellion. He arranged for a knotted cord to be passed to each of the Pueblo villages as a way of counting down to the day for the well-planned uprising—August 13, 1680. Upon learning the Spaniards had gotten wind of the plot, Pope called for an earlier attack—before dawn on August 10. The revolt quickly spread as Jicarilla Apaches and other natives joined the rebellion. The dissidents tied twenty-one friars to their altars and shot them with arrows.

The Spaniards fled, leaving many horses behind. The Puebloans' acquisition of this windfall speeded the animals' dispersion, increased the natives' mobility, and changed profoundly the nature of warfare in the American Southwest. Pueblo recapture of the Rio Grande valley was short-lived, however; within twenty years the Spaniards dominated the natives once again. Many Puebloans took refuge with Navajos. More important than the rebellion's temporary delay of the Spanish conquest were its long-range consequences: the dispersion of the horse and the spread of the Puebloan culture to the Navajos. [82]

Puett Reservoir State Wildlife Area. When full, this reservoir in Montezuma County has 140 surface acres at 7,261 feet elevation; it is surrounded by farmland with piñon-juniper woods, rabbit brush, and sagebrush. Managed by the Colorado Division of Wildlife, it gets irrigation-diverted water from Lost Canyon Creek. A conservation pool maintains the reservoir's fishery of walleye, northern pike, yellow perch, and channel catfish. From Mancos, Puett Reservoir is ten miles northwest on Colorado 184, a mile south on Montezuma 33, and a mile east on an access trail. [250]

quaking aspen. As the most widespread tree in North America, the "quackie" lends a special beauty to the upper part of the basin where it does its best to substitute for the colors of the eastern hardwoods. Also known as the western trembling aspen, its leaves pivot and flutter at the slightest breeze on their long, flat stalks, turned at right angles to the surface of the leaves. In its favorite elevation zone—6,500 to 9,000 feet—the aspen springs forth after forest fires or clear-cutting to shelter conifer seedlings. The evergreens then mature and reclaim their domain. The aspen's roots await the next fire that, perhaps decades later, will let them sucker again. Thus does the cycle continue. Bears like aspens and leave claw marks high on the trunks. For the beaver, the tree is a source of both food and building materials. That is why, in their search for furs, the nineteenth-century trappers sought out the aspen groves.

Once regarded as weed trees, aspens are now split into match sticks, shaved to make excelsior, chipped up for wood paneling, and mixed into spruce pulp for paper. But while its commercial uses have multiplied, many see its less tangible value, such as the habitat it affords. In the West no tree's beauty surpasses the golden autumnal foliage of the quaking aspen. [102, 233]

railroads. As in the greater West, the role of the railroads in the development of the San Juan Basin can hardly be exaggerated.

The Denver and Rio Grande (D&RG) came first to the basin by way of Monero. With its line to Silverton, completed in 1882, and its branch from Durango to Farmington, built in 1905, it became the most important means of transportation in the basin. As well as serving the mining, timber, cattle, and agricultural industries, it connected the basin to the national railroad system.

Also important was the Rio Grande Southern (RGS). Its rails went from the D&RG at Durango through the Rico mining districts to reconnect with the D&RG north of the basin. Several smaller lines, such as the three "baby" lines out of Silverton, fed the main roads.

Spurs, financed either by the railroad companies or their customers, also branched off the main lines to serve mining and lumber operations. Such was the Perins Peak Railway from Franklin Junction two miles west of Durango that ran up Lightner Creek to the Boston Coal Mine.

Here is a summary of the basin's approximate railroad mileage (excluding spur lines):

Main Lines

D&RG, Monero to Silverton	135
RGS, Durango to Lizard Head Pass	103
D&RG, Durango to Farmington	49

Feeder and Special-Purpose Lines

Dolores Lumber Railroads	60
Rio Grande, Pagosa and Northern	20
Rio Grande and Pagosa Springs	33
Rio Grande and Southwestern	10
Silverton	18
Silverton, Gladstone and Northerly	7
Silverton Northern	13
Total Basin Railroad Mileage	**448**

[63, 92, 100, 109, 176]

Red Apple Flyer. The train that ran on the Denver and Rio Grande Railroad between Durango and Farmington acquired this nickname, perhaps because it hauled fruit from Farmington. The railroad built the Farmington branch in 1905 and shut it down in 1969. [6]

red light district. The area of town where brothels were allowed to operate, as they were in Silverton, Durango, and other infant frontier communities. These sections are frequently called "tenderloin" districts after a notorious part of New York City. By some accounts, red lights were first linked to prostitution at Dodge City, Kansas. [223]

Redmesa, La Plata County, Colorado. Mormons migrated from New Mexico Territory and settled here in the early 1900s. They tried to make a go of farming through various La Plata River irrigation schemes, as do farmers today.

Redmesa (elev. 7,000 feet) is fifteen miles south of Hesperus on Colorado 140. [48]

Red Mountain Pass. This entrance to the San Juan Basin from the north separates the northern Mineral Creek drainage from south-sloping Red Mountain Creek. Nearby Red Mountain gets its name from the red iron oxide you see on its side. The pass is the scene of many snowslides on the highway from Silverton to Ouray. The East River Slide at the pass is notorious for burying winter travelers.

Otto Mears built a tollroad over the pass in 1882–1883 and followed it with his Silverton or "Red Mountain" Railroad a few years later.

Red Mountain Pass (elev. 10,018 feet) is nine miles north of Silverton on U.S. 550. [219]

Red Mountain Railroad. See **Silverton Railroad.**

redskin. This term for a Native American is not only derogatory but also inaccurate. Native complexions vary from white to quite dark. The skins of many are no more reddish than are those of Anglos or Africans. Some historians believe the term came not from the tinge of some natives' skin but from the color of their war paint. [118, 220]

Remington, Frederic (1861–1909). This artist, commercial illustrator, sculptor, and writer started his western career sketching scenes around Ignacio. Born in Canton, New York, the only son in an affluent family, Remington enrolled at the Yale School of Fine Arts, but his main interest was football. He joined the varsity to satisfy his urge for action. When his father died, Remington felt free to follow his delinquent tendencies and headed for Kansas to take up sheep ranching. After returning to the East to study at the Art Students' League in New York City, he headed again to the West and became engrossed with its crude violence.

His paintings, which number in the thousands, capture the mood of his subjects with unsentimental boldness; his bronze sculptures display vigorous movement. Many of his artistic works show the West as man's confrontation with a hostile physical environment.

Although best known for his artistry, Remington was also a successful journalist and fiction writer. His fiction works include *Pony Tracks* (New York: Harper's, 1895) and *Crooked Trails* (New York: Harper's, 1906). He capped his career in the Spanish-American War as a correspondent and artist before he died in Ridgefield, Connecticut. [124, 193, 217, 242]

Rico, Dolores County, Colorado. Once the world's largest silver camp, Rico earned a post office designation in 1879 and was Dolores County's first seat of government. Its 1879 silver boom was so explosive that a hundred cabins and dozens of commercial structures went up in one month. Because the town relied almost entirely on silver, with no base metals to fall back on, the 1893 collapse of silver prices devastated the community. The bursting camp (by some estimates, its population topped 10,000) became a ghost town of less than 200. Despite the mining industry's struggle to keep the community alive, the county seat was moved to Dove Creek in 1944, punctuating the town's demise.

Rico (elev. 8,827 feet) is forty-three miles northeast of Dolores on Colorado 145. [40, 65, 208]

Rio Arriba County, New Mexico. Now confined to an area in the northern part of the state, this county once swept from central New Mexico to California. After Congress divided the territory to form Arizona Territory, Rio Arriba absorbed part of Taos County. In 1887 the territorial legislature split Rio Arriba County to form San Juan County.

Rio Arriba, an original county of New Mexico Territory, was set up in 1850. County seat: Tierra Amarilla; population in San Juan Basin: 3,000 (est.); area within basin: 2,920 square miles (est.). [36]

Rio Grande. Although its watershed lies east of the Continental Divide, this river played an important part in the settlement of the basin. Snow melting on Stony Pass at the divide five miles east of Silverton might flow west into Cunningham Creek and to the Colorado River. Falling a few inches farther east, it could find its way to the headwaters of the Rio Grande.

Native Americans irrigated their fields with the river's water a thousand years ago, long before Coronado came north in 1540. From its mouth where you now find Brownsville, Texas, it served as a conduit for Spanish migration into New Mexico. In the lower reaches (Rio Abajo), the arid climate limited Spanish and Mexican use of the region to this river's valleys and those of its tributaries. The climate of its upper regions (Rio Arriba) was less severe, and settlers could wander farther from the streams. Many did so and crossed the Continental Divide into the basin.

Starting in 1915 with Elephant Butte Reservoir near Truth or Consequences, New Mexico, the Rio Grande's waters have been siphoned off for agricultural, municipal, and industrial use. Like the San Juan, the Colorado, and the other principal rivers of the West, it is no longer a natural flowing stream, but a plumbing system of dams, reservoirs, and canals designed to meet the demands of the growing West. [28, 80]

Rio Grande and Pagosa Springs Railroad. The New Mexico Lumber Company formed this line in 1895 to bring timber from north of Lumberton east to its mills at Chama. By 1911 the lumbermen had taken it thirty-three miles north, almost to Pagosa Springs. The line shut down when the timber industry faded in 1914. [63, 92]

Rio Grande and Southwestern Railroad. This railroad went south from Lumberton to the El Vado lumber camp on the other side of the Continental Divide; about ten miles of it was in the basin. Operated

Hispanic settlers, ca. 1890. Many Hispanics migrated to the basin from the Rio Grande valley before Anglo prospectors, cattlemen, and homesteaders advanced from the north and east. Courtesy Aztec Museum Pioneer Village.

through a pact between lumbermen and the Denver and Rio Grande Railroad from 1903 to 1924, it was essentially a timber-hauling operation. [63]

Rio Grande, Pagosa and Northern Railroad. Built by lumber operators in 1900, this line branched off the Denver and Rio Grande Railroad at Pagosa Junction and went twenty miles to Pagosa Springs. It became a branch of the Denver and Rio Grande in 1908. [63]

Rio Grande Southern Railroad. This line from Durango to Rico and points north was started by Otto Mears in 1890 for lumber, mining, and general cargo. Because some engineers believed there was no way to build a railroad through the labyrinth of mountains along its route, it was sometimes called "The Rio Grande Impossible."

The rails headed west out of Durango through Hesperus Pass to Hesperus, then followed the route along U.S. 160 and Colorado 184 to Dolores, ascending the Dolores River valley past Stoner and Rico. Curving up to Lizard Head Summit on its way to Ophir and the San Miguel River drainage, a series of switchbacks earned the nickname "Mears's Puzzle." Five branches served mines, including those near Mayday, Hesperus, and Rico. Its 163 miles of narrow-gauge rail (103 miles of it within the basin) crossed 130 bridges—one 95 feet high and another 554 feet long. So spectacular was the Ophir Loop section that it became one of the greatest wonders of railroad building in North America.

Less than two years after starting operations, however, the railroad got caught in the silver panic and went bankrupt. The Denver and Rio Grande took control of the line in the 1890s. Exposed to corporate manipulation under the control of the D&RG and ravished by floods, avalanches, and snow drifts, the road failed to achieve financial success.

Odd contraptions sometimes ran on the railroad. Fighting off ever-threatening financial disaster in the Great Depression, the operators rigged trucks and busses to run on the rails. One was a seven-ton 1928 Pierce-Arrow measuring eight feet wide and forty-three feet long. Another had a Wayne bus body and a GMC engine. From the honk of their horn and the swing of their gate on the wobbly tracks, they earned the disparaging nickname "Galloping Geese."

Defense needs, the lobbying of mining interest, and imaginative financing kept the line running through World War II. One scheme "sold" the RGS to the Defense Supplies Corporation through a lend-lease arrangement as if it were owned by a foreign country. (The railroad paid the money back.) Vanadium mining sustained the RGS through the postwar 1940s. The line quit operating in 1951. [65, 109, 175]

R

Rivera, Juan María de (ca. 1735–?). Leader of the first of three Spanish prospecting parties that journeyed from Santa Fe into the San Juan Basin during the period 1765–1775. Operating under the governor of New Mexico, Rivera made silver and gold discoveries along the basin's rivers, then went through the La Plata Mountains to the Dolores River. His travels charted the way for the Old Spanish Trail and the Domínguez and Escalante Expedition. [99]

rivers. From the tilt of the Continental Divide on its west and the slopes of the San Juan and La Plata Mountains to its north, the basin's watersheds drain to the Colorado River. The basin's mother river, the San Juan, swings south and west from Wolf Creek Pass. On the way to Four Corners and the Colorado, it collects water from the Pine and Piedra Rivers at Navajo Reservoir; Animas and La Plata Rivers at Farmington; and the Mancos River near Four Corners. A section of the basin's northwest drains to the Dolores. This river begins in the La Plata Range and starts south, but unlike the others, it courses north to enter Utah and meet the Colorado on its own. [104]

roadrunner. The state bird of New Mexico, this weak-winged cuckoo spends most of its time running around looking for insects, lizards, and small mammals. It is cream-colored and black with some blue iridescence on its long tail feathers. Roadrunners seldom fly more than a hundred yards and, as the name implies, would rather run down the road than take to the air. [223]

Rockwood, La Plata County, Colorado. A way station on the post road that ran from Silverton to Parrott City before the Denver and Rio Grande Railway went up the Animas Valley in 1882, this community continued to thrive after the coming of the railroad as passengers and freight transferred to stages and freight wagons bound for Rico. When the Rio Grande Southern Railroad provided an alternative to stage transportation between Durango and Rico in 1891, however, Rockwood faded. It remains a railroad stop.

Rockwood (elev. 7,376 feet) is on a subdivision-signed road that goes east off U.S. 550 fifteen miles north of Durango. [96, 101]

Rocky Mountain juniper. Like many of the West's white pioneers, basin residents often fail to distinguish this conifer from the eastern red cedar. Or they may call it the mountain red cedar—hence the name Cedar Hill for the juniper-covered knoll four miles south of the Colorado–New Mexico state line on U.S. 550. Juniper berries flavor gin, and the tree has a cedar fragrance, but its crooked, knotty trunks discourage commercial use. It grows at elevations of 4,500 to 8,500 feet, where you see it scattered across the semiarid hillsides, commonly mixed with piñon pine—hence the term "P-J forest." [233]

Rocky Mountain oysters. Beef testicles (originally buffalo testicles) that are prepared to be eaten, most commonly fried. Also called "prairie oysters" or "mountain oysters." [223]

Rocky Mountains. Of the four principal sections of this rugged mountain chain (Canadian, Northern, Central, and Southern), the Southern, where the basin rests, is the broadest. And although most of the chains 14,000-foot peaks lie elsewhere, the Southern section has the highest overall elevation. The San Juan belt of the Rockies forms a barrier to the northeast of the basin, as does the Continental Divide to the basin's east. The Rockies are geologically mixed, and nowhere is their complexity more evident than in the basin's topographic labyrinth.

rodeo. An exhibition or variety of sporting event that stems from cattle raising. Rodeo is a corruption of the Spanish rodeur, to "surround" or "round up." The term is appropriate, for the annual cattle roundup tests the cowboy's skills, especially so on the frontier before fences confined the livestock.

After the hands completed their cattle sorting and branding and drove the stock to market, they liked to exhibit their skills. Informal contests became formal competitions with rules and financial rewards. Modern rodeos include saddle bronc riding, bareback riding, calf roping, bull riding, and bulldogging (steer throwing). Women's barrel races, a horseback event in which the contestants ride a course marked with barrels, and cattle cutting, the separation of a cow from its herd by a horse and rider, are often secondary attractions. Some basin rodeos feature Native American cowboys. [217]

Rogers, Will (1879–1935). Humorist, actor, and commentator who paid an unannounced visit to Durango in July 1935, a month before he died in an airplane crash with celebrated aviator Wiley Post at Point Barrow, Alaska.

William Penn Adair Rogers started in vaudeville in 1905 doing rope tricks. He added humor to his act by poking fun at well-known personages, especially politicians, then became a movie actor and syndicated columnist. He's the man who said, "All I know is just what I read in the papers." [12, 133]

roundup. In the 1860s, when there were few herds on the open range, a cattleman could easily gather his animals in the spring, brand them, and put them back on their own ranges. He herded them again in the fall to select animals for market. When outfits multiplied in the 1870s and the chaos of mixed-up cattle got to be too much for the individual stockman, he got together with other growers for the spring roundup.

This, and the need to reduce cattle theft, gave birth to stockmen's associations that wrote roundup rules. These methods were normally put into state laws. Colorado's statute, for example, divided the state into sixteen roundup districts. In each district the governor appointed a board to guide the roundup program and hire a foreman—at three dollars a day. The county commissioners sometimes served as the roundup board.

On the date set for the roundup, you could see dozens of wagons pull into a camp. Each outfit provided its prescribed number of cowboys. The more experienced cut the cattle into herds by ownership. By recognizing conformation, head shape, twist of the horns, or behavior, some of hands could spot their owner's cattle without reading the brands. After each day's roundup the camp moved five or ten miles, taking with it thousands of cattle, hundreds of horses, and its fleet of supply wagons. At the next stop the cowboys roped fresh horses, saddled up, made the next loop, and endured another dawn-to-dusk day of hot and dusty cattle sorting.

By the turn of the century fences had divided the ranges, so roundups were no longer necessary. But you may still admire the teamwork of a trained cutting horse and his rider at many stock shows and rodeos. This and other roundup traditions carry on to remind us of the cow-country traditions of the nineteenth-century West. [54]

Russian thistle. See **tumbleweed.**

rustling. The origin of this euphemism for cow stealing is unknown. Some speculate that a journalist first used the term to make his writing more colorful.

Steal money and you're a thief. Steal a horse and you're a horse thief. But why, if you steal a cow, are you just a "rustler?" The answer: at one time so many drovers took cattle not necessarily their own that the practice was all but accepted. After the Civil War the Texas longhorns were so numerous that they had little value. And a lost calf on the open range was a temptation for even the honest cowboy. Even today, taking a cow worth hundreds of dollars seems somehow less serious than stealing the same amount in cash. Yet, to cattle growers, the loss is just as serious—more so if they lose a favorite breeding animal. [43, 54, 108]

Hiram W. Cox (center) and his five sons take time out from roundup chores on their Five Mile Mesa spread near Bondad, ca. 1880. When the Coxes and the Trubys fought over pastures, the mesa was called "Trouble Mesa." Courtesy Aztec Museum Pioneer Village.

Salmon Ruin. A large Chacoan apartment complex built during the eleventh century. Like many places where ancestral Puebloans lived, this is a multiple site; it developed in stages. The occupants who built the original 150-room complex in 1088–1090 were tied to the inhabitants of Chaco Canyon some fifty miles to the south. They left the site about 1130, but a few indigenous San Juan River valley residents used the place until 1185. Then a group of secondary occupants remodeled the structure by subdividing rooms and building more kivas. These people, who were culturally attached to the occupants of Mesa Verde some sixty miles to the northwest, abandoned the place in 1285, a century after they arrived.

The massive stone masonry of the ruin is a classic example of pre-Columbian Puebloan architecture. Its occupants were food growers and gatherers. As with other ancestral Puebloans throughout the Southwest, why they came when they did, and why they left, is a subject for ongoing archaeological study.

Heritage Park at the site tells about the cultures of other people who have lived in the Four Corners area and features a Pleistocene playa and related artifacts, a Puebloan Basketmaker pit house, Navajo hogans, Jicarilla Apache and Ute wickiups and teepees, and an old trading post.

George Salmon, who homesteaded the property in the late 1800s and for whom the ruin is named, protected the place from pot hunters and vandals for more than ninety years. San Juan County bought the twenty-two-acre site in 1969. Salmon Ruin and Heritage Park, including the San Juan County Archaeological Research Center and Library, is operated by the San Juan County Museum Association. It is two miles west of Bloomfield on U.S. 64. [147]

saloon. The public bar of the frontier did more than dispense drinks; it met other social needs as well. Since the basin had no end-of-the-trail cow towns, its barrooms were more akin to those of the mining camps. In the chaos that exploded around a gold or silver strike, one of the first tents to go up sold alcohol by the drink. Such emporiums evolved with the community, taking their place with the general store and the pharmacy along the main street.

Entering such a joint, you were likely to see a long bar down one side, backed by carved woodwork framing a large mirror. Opposite the bar a row of tables awaited you for gambling or group conviviality. From the back you may have heard the plink of a piano or the clack of billiard balls.

Why would you have gone there? In a mining camp you sought relief from loneliness and the drudgery of the hammer and drill, news from back east. New in town, you hoped for talk about a strike, a way to earn your next meal, or a place to sleep. As a steady customer, you could ask the saloonkeeper for a loan. As the camp grew, you might go there to join a vigilance committee, elect a governing board, or pick up your mail. Mostly you looked for entertainment. You found it in the form of a lottery, a prizefight, or a dance.

As a town matured, the general store, post office, church building, and boardinghouse assumed the accessory functions of the saloon and relegated it to the role of providing companionship and entertainment. But the saloon didn't need accessory activities to attract customers. Such was the demand for its normal functions that in Silverton, in the 1880s, thirty-four such emporiums operated. [49, 128]

Sandoval County, New Mexico. Much of this county is covered by the Jicarilla Apache Reservation. This was Chaco territory during the eleventh century. Encouraged by eighteenth-century Puebloan exiles from Spanish oppression along the Rio Grande, Navajos built defensive watchtowers and breastworks around their hogans.

Sandoval County was split from the northern part of Bernalillo County in 1903. County seat: Bernalillo; population in San Juan Basin: 200 (est.); area in basin: 450 square miles (est.). [87]

San Juan Basin. Many use this term; few define it (see Introduction). The region lies east of Utah and Arizona, north and south of the Colorado–New Mexico border.

To get to the basin's largest city, Farmington, here are the distances you would travel from various locations around the United States:

Durango saloon, ca. 1890. Mining camp saloons were a place to learn about a gold strike, find a job, pick up the mail, or lose your pay. Courtesy Fort Lewis College, Center of Southwest Studies.

Wichita, Kansas	700
San Bernardino, California	734
Walla Walla, Washington	1,012
Lake Charles, Louisiana	1,207
International Falls, Minnesota	1,579
Homestead, Florida	2,169
Cambridge, Massachusetts	2,334
Fairbanks, Alaska	3,371

[212, 253]

San Juan Basin Area Vocational-Technical School. As the agent for postsecondary vocational education in Southwest Colorado, this public institution serves Archuleta, Dolores, La Plata, Montezuma, and San Juan Counties. In cooperation with Pueblo Community College's Southwest Center, it offers associate degrees in the arts, sciences, and general studies. The school's campus is seven miles east of Cortez on U.S. 160. [255]

San Juan Basin Research Center. This agricultural experiment station's pioneering work in genetics has helped improve beef production throughout the world. Operating at 7,300 feet elevation, the station also works to enhance high-altitude range management. Since it has only a 100-day frost-free growing season, its crops are limited to small grains and forage grasses.

Its 6,300-acre site once served as the location for Fort Lewis, a military post that was moved here from Pagosa Springs in 1881 to subdue Native Americans. When the site became excess federal property in 1911, Colorado obtained the property for use as a school of agriculture. The school evolved to become Fort Lewis College. When the college moved to Durango in 1956, its abandoned site became a research station of Colorado State University, Fort Collins, as it is today. The center is four miles south of Hesperus on Colorado 140. [252]

San Juan College. Located in Farmington, with branches at Kirtland and Aztec, this school offers occupational certificates as well as associate degrees in the arts, sciences, applied sciences, general studies, and nursing.

San Juan Counties. Two of the four counties with the name San Juan lie within the San Juan Basin—one in each of its two states, Colorado and New Mexico. The others are in Utah and Washington.

San Juan County, Colorado. The first Anglo prospecting of present-day San Juan County occurred in 1860 and started rumors that the mountains were rich in minerals. The rumors turned out to be true, but development came slowly, delayed by the Civil War and Ute resistance to white intrusion. Not until 1873 did the Utes (under pressure) agree to surrender control of the mining areas. With the threat of hostilities gone, the miners still faced the hardships brought on by the isolation, harsh terrain, and high elevation of the San Juan Mountains in addition to treacherous wagon roads, vicious winters, a high cost of living, and a lack of money.

Even so, the prospectors came and the mining camps sprang up: Eureka, Howardsville, Animas Forks, Gladstone, Ironton, Red Mountain, Silverton. Then, in 1882, up the Animas River from Durango came the Denver and Rio Grande Railway, bringing food and coal and halving the cost of shipping ore for processing.

Overproduction and Congress's 1893 repeal of the Sherman Silver Purchase Act brought on a collapse in the price of silver. But the region's mining was not for silver alone. Gold, copper, lead, and zinc sustained it through the first two decades of the twentieth century. After that, mining was a transient visitor until 1994 when the last mine, the Sunnyside gold mine, shut.

Instead of coal and ore, the D&RG's successor, the Durango and Silverton Narrow Gauge Railroad, now hauls tourists into the county every summer where they provide a flurry of economic activity. But tourism, even helped by leftover mining cleanup, can sustain only a fraction of the population that prospered in the county's heyday of the 1880s.

San Juan County was split from La Plata County and established in 1876. County seat: Silverton; population (1990): 745; area: 392 square miles. [113, 115, 208]

San Juan County, New Mexico. Before the Spaniards arrived, the Navajos had a name for the country now encompassed by this county tucked into the northwest corner of the state. They called it Tohta, meaning "Among the Waters" or "Three Rivers," referring to the San Juan, mother of the basin's rivers, the Animas, and the La Plata—the streams that come together within its borders.

Unlike the Colorado county that bears the same name, this southern basin county saw no gold rush. And no Spanish or Mexican land grants stimulated its settlement. A few Hispanics drifted in before 1848 while it was still part of Mexico, but Mexico's rulers, like their Spanish predecessors, apparently failed to encouraged permanent settlement. The closest land grant was the Tierra Amarilla, forty-six miles to the east.

In the 1860s and 1870s prospectors came to the San Juan and La Plata Mountains and their south-sloping valleys. But as late as 1874, there were so few white pioneers in northwestern New Mexico that President Ulysses S. Grant met little opposition when he ordered a hunk next to Colorado set aside as a Jicarilla Apache Reservation. (The Apaches showed little interest in using the territory because it was subject to raids by roving bands of Utes and Navajos.)

In 1875 rumors circulated that Grant would reopen that part of the basin below the Colorado–New Mexico state line for settlement. Homesteaders, weary of not finding sudden riches in the Colorado mountains, came south into New Mexico. Among them was William P. Hendrickson, who persuaded four men to go with him to the lower valley. They settled between the rivers where the Animas meets the San Juan. Like many white pioneers, they didn't wait for the government to declare the country open for their entry. The settlers saw how vast stretches of rich bottomland and sandy loam on the lower mesas could be irrigated. As they uncovered evidence of ancient cultivations, they perhaps realized that some mysterious inhabitants had used that idea and farmed there hundreds of years before. And standing on a mesa, they may have observed also how the valley's buffalo and grama grasses could feed herds of livestock.

After the area became open for entry in 1880, it didn't take long for hundreds of other pioneers to follow Hendrickson's footsteps. By 1892 the settlers had planted perhaps 70,000 fruit trees. That year's county fair featured peaches with nine-inch circumferences and apples weighing nineteen ounces. The valley's fruit took the sweepstakes at the territorial fair in Albuquerque. The county's agronomists capitalized on their fame by exporting tons of fruit; in a year they produced 300 tons of apples and over 200 tons of peaches as well as pears, plums, cherries, and various varieties of berries. With their grapes they made 1,800 sixty-cent gallons of wine. The county was still exporting fruit by the carload when the Denver and Rio Grand Railroad brought its line south from Durango in 1905. The train became known as the "Red Apple Flyer."

The big news item of 1922 was a harbinger of developments that would dramatically change the county's economy, lifestyle, and character—the Midwest Refining Company brought in a 350-barrel oil well on Navajo-leased land fifteen miles west of Farmington. Ever since, extracting, moving, and refining oil has been a principal part of the county's economy. When the nation's demand for oil boomed on the 1950s, the D&RG's focus changed from exporting fruit to importing pipe and oil-well pumps. Farmington's population increased fivefold in a single decade as subdivisions obliterated the orchards. Orchard Avenue, in the county's exploding Farmington metropolis, became an anachronism.

San Juan County was split from Rio Arriba County and established in 1887. County seat: Aztec; population (1990): 91,605; area: 5,942 square miles. [15, 51, 85, 204]

San Juan Generating Station. With coal from the La Plata and San Juan Mines, this 1,700-megawatt power plant is operated jointly by the Public Service Company of New Mexico and the Tucson Electric Power Company. The road to the station goes north off U.S. 64 nine miles east of Shiprock. [146]

San Juan Mine. This mine feeds coal to the nearby San Juan Generating Station. A road north off U.S. 64 nine miles east of Shiprock provides access to the mine.

San Juan Mountains. This group of the Rocky Mountains extends westward from Colorado's San Louis Valley nearly to the Dolores River. It is limited on the north by the Gunnison River and on the south by the foothills draining to the San Juan River. Within the San Juans lie the La Plata and Needle Mountain Ranges. [244]

San Juan National Forest. Covering over 3,000 square miles on the west slope of the Continental Divide, this is one of 256 sites managed by the Forest Service. Running 120 miles from east to west and 60 miles from north to south, it stretches through seven counties and encompasses the Weminuche Wilderness and parts of the South San Juan and Lizard Head Wildernesses. Administered as part of the San Juan–Rio Grande National Forest, the San Juan portion has ranger stations at Dolores, Mancos, Durango, Bayfield, and Pagosa Springs.

The area features a collection of lakes, canyons, cataracts, waterfalls, and a broad spectrum of elevations and vegetative zones. Its Needle Mountains are among the roughest ranges in the United States. In addition to its natural attractions, the area has pre-Columbian sites at Chimney Rock, a Nordic ski area north of Durango, alpine ski trails near Mancos, reservoirs (McPhee, Vallecito, and Lemon), and numerous campgrounds and picnic sites. [160]

San Juan River. The snows give birth to this, the basin's mother river, near Wolf Creek Pass. It tumbles through rocky canyons, broadens as it approaches Pagosa Springs, then becomes a desert river before leaving the basin near Four Corners, ninety miles from its source. Along the way, it gathers most of the water that falls on the basin. Navajo Reservoir, at the Colorado–New Mexico border near Arboles, interrupts the river's flow. About a hundred miles after leaving Colorado and entering Utah, the San Juan used to crash into the Colorado River. Now it drifts unceremoniously into the backwash of Lake Powell. [104]

San Juan River Highway Bridge. Touted as "the longest functioning historic steel overhead truss highway bridge in New Mexico," this 1936 bridge features six 166-foot spans. It takes U.S. 666 and U.S. 64 over the San Juan River south of Shiprock. [238]

San Juan Skyway. A route designated as a byway by the Colorado Scenic and Historic Byways Commission whose basin portion extends from Lizard Head Pass south to Cortez, then east to Durango and north to Red Mountain Pass. [209]

Sanostee, San Juan County, New Mexico. Frank Noel and C. H. Algert bought this trading post in 1905 and gave it its name. The post was built about 1900. Sanostee is ten miles north of Newcomb on U.S. 666 and ten miles west on Navajo 34. [87]

Santa Fe Trail. Of the famous trails that traders and pioneers followed west, the Santa Fe is the most significant to the basin. At its western terminus, for which it was named, it met the Royal Road (El Camino Real in Spanish), which led to Mexico City, and the Old Spanish Trail, a route of commerce through the basin to California.

Old Spain forbade exchange with foreigners, but when the New World Spaniards split from their European founders in 1821 and opened the country for trade, pack trains of merchandise headed for the region. Trader William Becknell initiated the Santa Fe Trail that year by hoofing 900 miles from Franklin, Missouri, to exchange his manufactured merchandise for silver and furs. When freight wagons replaced pack trains, they made some fifteen miles a day and took four months to complete the round trip.

From Franklin, Westport, and Independence in Missouri and Fort Leavenworth in Kansas, the trail headed west through Council Grove, Fort Larned, and Dodge City before splitting at the Cimarron Cutoff. From there the arid southern route cut through what is now the tip of Oklahoma's panhandle before continuing southwest and over Glorieta Pass to Santa Fe. The less hazardous, but longer, northern leg skirted through Bent's Fort and over Raton Pass in modern-day Colorado before branching to Taos or rejoining the main trail north of Las Vegas, New Mexico.

In its heyday the trail's transport businesses employed some 10,000 men driving 6,000 mules and 28,000 oxen pulling 3,000 freight wagons. For nearly six decades, until the railroads took its business in 1880, the Santa Fe Trail served as the principal road for commerce and western migration to New Mexico. This economic and cultural link not only provided a southern connecting point for pioneers; it started the "gringo" invasion, the takeover of the Southwest by the United States. [28]

Sapiah (1840–1936). At Ouray's request, this Capote Ute tribal subchief, also known as Buckskin Charlie, became leader of the Southern Utes shortly before Ouray's death in 1880. As a chief after Ute subjugation, he played an important role in his tribe's adjustment to reservation life. With 350 other Utes he paraded at President Theodore Roosevelt's inauguration in 1905. [74, 119]

Scenic and Historic Byways. The Colorado Scenic and Historic Byways Commission's blue columbines mark the way of routes that identify and interpret many of the basin's scenic, historic, cultural, and recreational features. The Alpine Loop goes east out of Silverton on Colorado 110. Lying entirely within the basin, the Trail of the Ancients circles from Four Corners through Cortez, Dolores, and Pleasant View, then heads west to Utah. The San Juan Skyway passes through Silverton on its way south to Durango, continues west to Cortez, and cuts northeast through Rico before leaving the basin at Lizard Head Pass. [209]

Sheek, James Lorenz (1864–1949). This Mancos cattleman brought to the basin his experience with the famous drover Charles Goodnight. His father, John Wesley "Wes" Sheek, was Goodnight's step-brother and partner until Goodnight decided to drive cattle north from Texas. James Sheek trailed herds with Goodnight into New Mexico and Colorado and later helped Goodnight breed Angus cattle with buffalo to produce "cattalo."

After starting his Mancos Valley cattle business in 1899, Sheek brought his family by covered wagon from Oklahoma five years later. He held one of the basin's earliest forest-grazing permits. [53, 59]

sheep. Historians and writers have left the story of sheep in the American West largely untold. While many exploit the conflicts between cattlemen and sheepherders, few give the sheep industry a fraction of the space devoted to cattle. It is easy to see why.

The cowboy cuts a romantic figure astride his steed, lording over his longhorns. The sheepherder goes afoot. Although sheep are respected in other countries (New Zealand is an example), only the most devoted

Sapiah (Buckskin Charlie), ca. 1920. This subchief of the Capote band followed Chief Ouray as leader of the Southern Utes. Sapiah joined President Theodore Roosevelt's inaugural parade and enjoyed flying in barnstorming airplanes. Courtesy La Plata County Historical Society.

director can bring off a motion picture featuring ovines in an American Western.

Like cattle, the first sheep of the Southwest were of Spanish origin; Columbus unloaded some on his second voyage to Hispaniola in 1493. A half century later, sheep were among Coronado's provisions when he sought wealth in present-day New Mexico, Kansas, and Texas. Juan de Oñate brought sheep up the Rio Grande valley as far as Rio Chama, thirty miles east of the San Juan Basin. By the eighteenth century sheep were a source of meat and wool for both Native Americans and Spanish immigrants throughout New Mexico.

During the Southwest's Spanish period, sheep were its largest agricultural product; up to half a million head annually moved south to Mexico City. In the early nineteenth century it was not uncommon for a New Mexico "sheep king" to have a half million head spread over the province. By the time the United States wrested control of the region from Mexico in 1848, the industry had spread over the entire West.

While cattle drives stir our spirit of adventure, sheep drives were no less important. And some of the West's most famous pioneers took part. In 1852 Kit Carson and Lucien Maxwell drove two bands totaling 13,000 head from the Rio Grande valley to Wyoming. (A large cattle drive might have one-fourth that number.) The same year, Richens "Uncle Dick" Wootton left Taos with 9,000 head and crossed the Continental Divide above the Rio Grande's headwaters just east of the basin on his way to California's Sacramento Valley. His band following the hoofprints of 25,000 sheep driven to California's gold fields by Miguel A. Otero and Antonio José Luna in 1849.

As Texas cattle bred unchecked during the Civil War, so too did California sheep. After the conflict, bands of the woolies moved east to railheads. Many cattlemen have described the hardships their animals faced on the trail. Sheepmen faced those trials plus poisonous weeds, wolves, bobcats, coyotes, and eagles. As was the case with cattle, that part of the West lying northeast of the basin in Colorado underwent a varied sheep experience. Animals came from the east to mingle with the Spanish stock migrating north from Mexico. The basin's mountain barriers and isolation delayed sheep settlement until a relatively late date.

Bands on the highways between pastures and the bleating from passing stock trucks evidence the basin's continuous alliance with the ignoble sheep. [127, 204]

Sheep Springs, San Juan County, New Mexico. This trading post is one of a string that started up on the east slope of the Chuska Mountains after the Navajos were released from their four-year imprisonment near Fort Sumner.

Sheep Springs is forty-six miles south of Shiprock on U.S. 666. [93]

Sheridan, Philip Henry (1831–1888). This Civil War hero visited the basin during the height of his career. Like his sometime commander, William T. Sherman, Sheridan is most often remembered for his engagements against the Confederacy; yet he spent most of his military career, including his service before the Civil War, expanding the frontier.

After Sheridan got his education at the U.S. Military Academy and served at Fort Duncan, Texas, and Fort Reading, California, he commanded campaigns against the Yakimas and Cascades in the Pacific Northwest. Not until the Civil War was well underway did he get a chance to return east. As a commander in the Army of the Cumberland, he shared in the humiliating defeat at Chichamauga and the siege at Chattanooga, but his charge up Missionary Ridge won high praise. General Ulysses S. Grant promoted him to command the cavalry in the Army of the Potomac.

As commander of the Army of the Shenandoah, Sheridan drove the Confederates out of the Shenandoah Valley, then applied a scorched-earth policy to keep it from growing food for the Confederates. After achieving the rank of general in 1864, Sheridan won the Battle of Five Forks and forced the Confederate commander, Robert E. Lee, to abandon Richmond and withdraw to Appomattox.

The San Juan Basin came under his command in 1868 when Grant assigned him to command the Department of the Missouri—a military region comprising Illinois, Missouri, Kansas, Oklahoma, Colorado, and New Mexico. In his later capacity as commander of the Division of the Missouri (including the Department of the Missouri and four other departments stretching over a million square miles), he visited the basin.

During an 1879 tour of inspection that took him to the White River Agency (less than four months before the Meeker Massacre at that location) and Ouray, he rode over the mountains to Silverton where cheering crowds greeted him. He continued down the Animas Valley to Pinkerton Springs, near the site of the original Animas City. At Pinkerton's homestead the forty-eight-year-old general admitted that, after sitting astride a horse for 700 miles, he was too exhausted to ride farther. He consented to resume the trip, at least to the Shaw House at Animas City, in an army ambulance. There he rested before his cavalcade continued to Fort Lewis, then located near Pagosa Springs.

During his tour of the basin, Sheridan saw that Fort Lewis should be moved to a more westerly location. A few months later the fort's troops were called out to protect the Animas Valley from a perceived Ute threat. The fort was moved permanently to the La Plata River valley.

S

Sheridan was the nation's fighter of Native Americans from 1867 until he followed Sherman as commander of the entire army in 1884. During this period his troops fought 619 engagements and suffered 1,256 casualties, including 565 who died. (There is no definitive record of enemy casualties.)

Even less than his direction of the Indian Wars do historians note his efforts at preserving a national jewel, Yellowstone Park. He first became acquainted with its marvels during an 1870 inspection tour. What he saw during an 1882 trip enraged him: hide hunters were slaughtering the wildlife. For two dollars a day, an arm of the Northern Pacific Railroad had monopoly rights to develop the park. Sheridan threatened to protect the park with troops, then rallied political friends to the park's defense. As a result, Congress enacted laws protecting and expanding Yellowstone.

Before he died in 1888, Sheridan was privileged to wear four stars by joining George Washington, Ulysses S. Grant, and William T. Sherman in holding the rank of general of the Army of the United States. [67, 197, 217]

Sherman, William Tecumseh (1820–1891). Although best known for his Georgia campaign that split the Confederacy and helped assure Union victory, this U.S. general climaxed his career by serving in the West. As commander of the Division of the Missouri, which encompassed the San Juan Basin, and as the general of the Army of the United States under President Ulysses S. Grant, Sherman helped bring peace to half a continent.

Sherman graduated from the U.S. Military Academy but left the army in 1853. He was president of what is now Louisiana State University in 1861 when Louisiana seceded from the Union. As a native of Ohio, Sherman left the South and joined the Union army. After fighting at Bull Run, Shiloh, and Vicksburg, he became commander of the Army of Tennessee in 1863 to lead the Battle of Chattanooga. As supreme commander of the armies of the West in 1864, he captured Atlanta and made what became known as "Sherman's March" to Savannah, laying waste to the intervening territory.

Perhaps Sherman's guidance of the enormous and complex job of protecting settlers scattered over thousands of miles of hostile territory is largely overlooked because during the so-called Indian Wars, the nation was technically at peace. In the West he confronted no giant armies. The Native Americans stubbornly hung on to their land, but they had neither bugles nor cannons.

To protect the ever-advancing but widely scattered miners, cattlemen, homesteaders, and other settlers from marauding defenders and

181

confine the natives to reservations was both a tremendous and thankless task. In addition to dealing with skilled fighters maneuvering rapidly over difficult terrain and great distances, Sherman found his army facing the Department of the Interior, peace societies, and a niggardly Congress. And in the Southwest he underwent all this frustration to make safe a land he thought no homeseeker in his right mind would want.

Sherman engaged the Confederates for four years. He led the vastly more complex job of subjugating Native Americans for eighteen. [23, 217]

Sherman Silver Purchase Act. Perhaps no act of Congress has had a greater impact on the San Juan Basin. By inflating the price of silver and then letting it collapse, the government put silver production in chaos.

All this happened back in the days of "real" money—gold and silver coins. You could send your paper money, called certificates, to the treasury and get gold and silver in return. Until the 1870s, the mint bought silver and gold at a ratio whereby sixteen ounces of silver had the same value as an ounce of gold. That's when western prospectors made big strikes, silver rolled out of the mines, supply went up, and market value went down. To get the best price, producers refrained from selling to lower bidders and sent their silver directly to the mint. But as supply went up, the mint's price also went down.

Congress tried to stabilize silver prices by tinkering with the monetary system, but the ratio kept falling until it took twenty-two ounces of silver to get an ounce of gold. Silver needed help.

Enter the Sherman Silver Purchase Act of 1890. Although it didn't set a price, the act required the treasury to buy 4.5 million ounces of silver a month—almost the nation's entire output. As the government bought more silver, prices rose, just like Congress wanted. Predictably, producers took still more silver to the mint, and they weren't satisfied to go home with paper certificates. They wanted gold. If this kept up, the treasury could run out of gold. Certificate holders would be left with low-value paper. The monetary system would collapse. Congress had to act.

Exit the Sherman Silver Purchase Act. After propping up the price of silver for three years, from 1890 to 1893, it was scuttled. With this turnabout, the value of silver toppled. It now took over thirty-two ounces of silver to get an ounce of gold. Silver was worth only half what is was before the boom began.

Mines closed, trains stopped running, booming mining camps became ghost towns. In the San Juan Basin, Rico, overly dependent on silver production, saw its population plummet from thousands to hundreds. Other mining towns where gold and base metals diversified the economy, like Silverton, fared better. But even they would never return to their heydays when silver was overproduced to reap the prices shored up by an act of Congress. [63, 208]

Shining Mountains. A Ute name for the spectacular mountains of the territory they once dominated. The clear, dry air above Colorado's Rocky Mountains contributes to their brilliance, or shine, even in late summer when their high valleys cradle little snow. [86]

Shiprock, San Juan County, New Mexico. Thomas Keam scouted this area for a Fort Defiance subagency location in 1872, but the station didn't become a reality until thirty-one years later. Its first superintendent, W. T. Shelton, gave the place a boost when he got the idea of inviting Navajos to assemble and display their crafts for area traders. The fairs, held the first week of October, proved to be a successful market for the Native American rugs and other crafts. The community remains a center of commerce for the Navajos.

Shiprock (1990 pop. 7,687, elev. 4,903 feet) is twenty-six miles west of Farmington on U.S. 64. [71]

Shiprock Pinnacle. On a clear day without power plant haze, you can see this schooner-shaped rock a hundred miles away. Rising 1,700 feet off the desert floor on the Navajo Reservation, it reaches a height of 7,178 feet elevation. Its appearance comes from solidified lava walls flanking igneous rock. To the Navajos, the tower is a sacred place and off limits to climbers. The pinnacle is eleven miles southwest of the town of Shiprock. [71, 142]

Silverton, San Juan County, Colorado. This county seat sits in the middle of Baker's Park, a deep, flat-bottomed cavity tucked into a maze of nearly 14,000-foot mountains. Anglo-American prospectors found gold and silver in the San Juan Mountains around Silverton in the 1860s.

After the Utes surrendered the area for mining in 1873, a mining district boomed. By taking advantage of its location on flat ground to attract smelters and mercantile enterprises, the community quickly became the business center for the neighboring mining camps. It secured urban dominance when the Denver and Rio Grande Railway arrived in 1882. Three feeder lines (the Silverton, the Silverton Gladstone and Northern, and the Silverton Northern) soon branched up the valleys to serve the mines.

When silver's value plunged in 1893, silver production collapsed. Unlike mining areas that were too dependent on silver, however, the San Juan district had gold and base metals to fall back on. But by 1940 the

The schoonerlike Shiprock pinnacle rises 1,700 feet off the desert floor. Located on the Navajo Reservation near Shiprock, this sacred "rock with wings" is off-limits to climbers. Courtesy Aztec Museum Pioneer Village.

branch lines had stopped rolling, confirming the fate of both precious metal and base metal mining. The train still runs to Silverton, bringing tourists and their money to town in the summer, but efforts to rejuvenate the mining economy have brought only sputtering success. With the closing of the Sunnyside Mine in 1994, the mining industry was reduced to cleaning up the mess left by decades of environmental abuse.

Silverton (1990 pop. 716, elev. 9,318 feet) is forty-seven miles north of Durango on U.S. 550. [113, 115, 208]

Silverton, Gladstone and Northerly Railroad. Started in 1899, this was the shortest of three roads built out of Silverton to serve the mines. Otto Mears built the other two (the Silverton and the Silverton Northern) whereas mining investors built this one.

On its way seven miles up Cement Creek (Colorado 110) to the Gold King Mine, the track gained 1,300 feet in elevation. When the Gold King closed in 1909, Mears leased the line from the railroad's investors and combined it with his other two railroads. He bought it outright in 1915. Operations quit about 1924, but the rails were left intact in the hope that the Gold King would reopen. That didn't happen, so the rails came up in 1938. [39, 63, 110]

Silverton Northern Railroad. This line out of Silverton featured a contraption called a railroad bicycle—two bicycles, side by side, with their chains geared to the wheels of a homemade handcar. The railroad also had a 7 percent grade, a terminal at 11,200 feet elevation, and a Pullman car with dining accommodations.

Organized in the 1890s by Otto Mears, it ran thirteen miles up to Animas Forks. Like the other two short railroads out of the mining town (the Silverton and the Silverton, Gladstone and Northerly), it hauled coal and supplies to the mines and returned with ore for processing. Its last four miles to Animas Forks were so steep the engine pushed the cars for fear that if a pulled car became unhitched, it would overcome its brakes and take off like a loose cannon.

The Silverton Northern lasted until 1939, when the Sunnyside Mine experienced one of its several shutdowns; the rails came up in 1942. Colorado 110 and Forest Service 586 follow the abandoned railroad bed. [39, 63, 109]

Silverton Railroad. Sometimes called the "Red Mountain Railroad" because it went eighteen miles over Red Mountain Pass to Ironton in Ouray County, this was the first of three short railroads built out of Silverton to serve the mines. (The other two were the Silverton Northern and the Silverton, Gladstone and Northerly.)

At one gulch it was impossible to construct a curve sharp enough for the terrain. The engineers met the challenge by building a covered turntable at the end of the gorge. While the cars waited on an extension, railroad hands turned the engine around. With the cars once more in tow, the engine went back the way it had come, then veered off at a switch to proceed on the other side of the canyon.

Built in 1887–1889 by Otto Mears's Silverton Railroad Company on the bed of his famous Rainbow Route tollroad, the line went into receivership in 1898. It was sold under foreclosure in 1904 and reorganized by Mears as the Silverton Railway. The Silverton quit running in 1922, a victim of mining's decline; its rails came up in 1926. U.S. 550 follows the railroad's route. [39, 63, 109, 175]

Simon Canyon Recreation Area. This area below Navajo Reservoir's dam on the San Juan River features premier trout fishing. It is a facility of the Bureau of Land Management. [227]

Sleeping Ute Mountain. If you look southwest from the vicinity of Cortez, you will see a mountain skyline resembling the image of a man lying on his back—head to the north, face to the sky, arms folded across his chest. Some Utes say this formation is one of their gods. For some reason it became angry, gathered the rain clouds in its pockets, and lay down to sleep. But sometimes the clouds slip out and hang above the peaks. Then the rain comes. [50]

Southern Ute Reservation. A 480-square-mile territory reserved for members of the Mouache and Capote Ute bands and where band members received allotments of land from the United States. It stretches sixty-five miles along the north side of the Colorado–New Mexico border from Mesa Verde National Park to a point south of Pagosa Springs.

The entry of miners into Colorado after the gold strikes west of Cherry Creek (now Denver) in 1859 pressured the federal government to reduce the Utes' territory. The Treaty of 1868 removed the Utes from the Rocky Mountains' east slope and set up a reservation covering most of western Colorado. When the miners, many of whom migrated from the Denver area, entered the San Juan Mountains, the pattern repeated. The Brunot Agreement of 1873 took a large rectangle from the 1868 reserve.

At the same time, New Mexican settlers wanted the Utes removed from their territory. Congressional action moved New Mexico's Utes to the Colorado reservation in 1877. As you might expect, the Utes were not welcomed to southern Colorado. Several approaches to the problem under consideration fell victim to the Meeker Massacre of 1879.

The reservation, as you see it today, grew out of an 1895 act of Congress. Members of the Weminuche band opposed the allotment idea, so the west end of the reservation was set aside for them to hold in common; that part became the Ute Mountain Ute Reservation. After the members of the Mouache and Capote bands received their land allotments, the remaining land was opened for white settlement in 1899. It became known as the Ute Strip. [74, 86, 208]

Southern Ute Tribe. An association of Utes that controls much of the Southern Ute Reservation where 1,100 of its members live. As required by Congress, the tribe became a formal organization with a council and chairman in 1936. Headquartered at Ignacio, its people are largely descendants of the Mouache and Capote bands that hunted in southern Colorado and northern New Mexico—the southern portion of the Utes' native territory. [74, 119]

South San Juan Wilderness. This 200-square-mile area lies within the San Juan–Rio Grande National Forest astride the Continental Divide. Set up by Congress in 1980, the region is known for its dramatic waterfalls through gorges cut eons ago by giant ice fields. On a compass bearing, it is ten miles south of Wolf Creek Pass.

Spanish and Mexican land grants. Land rights granted by Spanish and Mexican rulers to reward their supporters and encourage settlement.

Spain, like many governments, secured empirical expansion by granting land to faithful citizens. Its rulers carried the practice to the New World. After New Spain declared its independence and formed the Republic of Mexico, Mexican governors continued the tradition. The descriptions of these grants came not through official surveys but by ambiguous traditions. To mark off land, you might pile up rocks around its perimeter or just walk around it throwing dirt in the air. Further, the authority for making such grants was often unclear, giving rise to uncertainty and fraud. With huge acreages at stake, favoritism, graft, political corruption, and legal shenanigans permeated the process. Under Spanish and Mexican governance, these land grants were the original and basic system for land distribution throughout California, New Mexico, and other portions of the western United States. By the Treaty of Guadalupe Hidalgo at the end of the Mexican-American War, the United States agreed to honor the grants within the territory given up by Mexico. Thus did the United States inherit a land title mess.

The San Juan Basin was not exempt from this experience. The history of the Tierra Amarilla grant, which slops over the Continental Divide into the basin from the Rio Grande valley, reeks of chicanery and political subterfuge. [44, 129]

squirrel. The basin's small pine squirrel, or chickaree, scampers around the woods gathering vast quantities of conifer seeds. Though entertaining, this rodent does not offer the attractions of the Abert's squirrel (or tassel-eared squirrel) that you commonly see in the ponderosa pine forests—with its fluffy tail, black or salt-and-pepper coat, and magnificent ears. [135]

stagecoach. See **Concord Coach.**

state parks and recreation areas. Both the Colorado Division of Parks and Recreation and the New Mexico Parks and Recreation Division maintain recreation facilities at Navajo Reservoir. Colorado's division also manages the Mancos State Recreation Area north of Mancos.

state trust lands. Lands that Congress granted to Colorado, New Mexico, and other states along with statehood. The idea behind the grants was to give infant states a source of revenue. States are required to manage many such lands (often called "school trust lands") to achieve revenues from rents, royalties, and land sales for the benefit of public schools. State land boards may open certain trust lands for specific public uses, such as wildlife-related recreation.

The Colorado Land Board has opened 640 acres in Disappointment Valley (Dolores County) and over 1,000 acres in the Weber Canyon and Menefee Peak areas near Mancos (Montezuma County). The Colorado Division of Wildlife manages the areas. [140]

Stockton, Ike (1852–1881). Outlaw head of the Stockton-Eskridge gang, he became known as the San Juan Basin's most notorious gunfighter. Whether he deserves that reputation is a matter for debate.

On a hillside plateau overlooking the Animas River, gravestones lean among the grama grass and squawbush to divulge a graveyard lying above the part of Durango that used to be Animas City. Face down within the graveyard, obscured by seasons of overgrowth, lies the tombstone of Isaac T. "Ike" Stockton, who died of a gunshot wound inflicted by a deputy sheriff September 26, 1881. (However, the stone reads "In Memory of Isaac Stockton, Born Feb. 20, 1852, Died Oct. 26, 1881." The discrepancy in the date waits for an explanation.)

Local historians view Stockton as Durango's claim to gunslinger fame, a personal example of the area's turbulent birth in the early 1880s. But notorious though Ike Stockton is by local lore, he lacks a wider standing. O'Neal's *Encyclopedia of Western Gunfighters* mentions him only as the older brother of Port Stockton. Likewise, Jessen's *Colorado Gunsmoke* gives Ike only brief mention. And although he spent several years riding in New Mexico Territory amid some of the most violent episodes of the West, Stanley's *Desperadoes of New Mexico* honors him with little more than a footnote.

Many personal accounts of life in the San Juan Basin refer to the Stockton-Eskridge gang as "dreaded" or "feared." Though he was, by all accounts, a bandit, he was apparently neither proud nor ashamed of it. He considered himself to be just a "normal" operative. In the only picture purported to be of Ike, he looks at us from a stiff pose dressed in the formal portrait attire of his era, one of four brothers. But which one is he? The names of the subjects have been obliterated, apparently reflecting the shame of his descendants. Why do accounts of his escapades refer to him as one of only two brothers? Did the others obscure their brotherhood because of Ike's reputation?

Isaac T. "Ike" Stockton came to the basin from what is now Cleburne in northeastern Texas—that part of the state known as "the violent triangle." He grew up there in fear of Comanches and thugs who operated with impunity in the absence of civil authority. In his youth he experienced, at least secondhand, the clash of the Civil War. Men went east, leaving behind children, women, and old men to fend for themselves. In their absence the young and the old formed home guards, ostensibly to fight an invading army, but used more to ward off deserters from both North and South.

Young Stockton lived amid the terror of radical reconstruction and later saw the excesses of cattle drovers on recess from the Shawnee trail. He migrated to Colfax County, New Mexico, and Las Animas County, Colorado, where a maze of conflicting land claims, born of the Beaubien-Miranda and other land grants, made gunplay a common occurrence. Twice, when his younger brother Port was arrested on suspicion of murder, Ike felt obliged to break him out of jail. Ike's path crossed Lincoln County, New Mexico, where within sight of Ike's saloon, Henry "Billy the Kid" McCarty and his cohorts dry-gulched Sheriff William Brady.

One of his first stops in the San Juan Basin was at ex-Lincoln County gunman George Coe's place. Perhaps an incident that occurred while he was visiting Coe sheds light on Stockton's omission from lists of western gunfighters. Another of the Coe clan, Lou, had a ranch next to some settlers' cultivated fields. When Lou's cattle invaded the plot and damaged their crops, the farmers demanded compensation. The dispute went to Justice of the Peace Court. George Coe and Stockton went to the hearing. When Judge Halford ordered the spectators to disarm,

Stockton and Coe expressed their displeasure and left the courtroom. The constable told them to disarm or leave the premises. When Stockton yelled a defiant retort and drew his pistol, the revolver's cylinder rod fell out!

He then dashed for the street to grab his Winchester from its saddle sheath. Meanwhile Coe outdrew the constable and defused the situation by verbally restraining Stockton. The mutterings of the short (five-foot, four-inch) stocky gunfighter standing by the hitching rail are left to your imagination.

After Port was gunned down by members of the Farmington Stockmen's Protective Association near Flora Vista, Ike continued his habit of taking up for his little—or at least younger—brother. In a gulch off La Plata Canyon north of Farmington, he tracked down and murdered one of the vigilantes, Aaron Barker. (That's why it's called Barker Arroyo.) When members of the association rode to Durango seeking Stockton, they started the basin's most famous gunfight as they shot it out—unsuccessfully—with the Stockton-Eskridge gang.

Stockton's death came not from an ambush or a barroom dispute, both common occurrences in the frontier West, but from his disregard for "honor among thieves" and his violation of the Code of the West. One night in August 1881 some of the Stockton gang—Stockton not among them—started a row at the bar of Silverton's Diamond Saloon. When Marshal D. C. "Clate" Ogsbury approached to make an arrest, gang member Bert Wilkinson drew his gun and shot the marshal, killing him instantly. He then rode off into the night, leaving the others behind. (One of those was "Black Kid" Thomas, whom the townspeople promptly lynched.) The outraged Silvertonians offered a $2,000 reward for Wilkinson's capture.

To collect the reward, Stockton arranged to meet his fellow gang member, got the drop on him, and turned him over to the La Plata Coun.y sheriff. Even those who viewed every quarrel as a private quarrel could no longer defend a gang leader who would turn on one of his own. Among those who now sought Stockton's arrest was La Plata County deputy sheriff James J. Sullivan.

On September 26, 1881, Sullivan, a New Mexico warrant for Stockton's arrest in his pocket, spotted Stockton near First Street and "H" Street (now Tenth Street and Main Avenue) in Durango. As Sullivan confronted Stockton, the lawman drew his gun. The pistol went off prematurely, and the bullet blasted open Stockton's thigh. Witnesses carried him to the office of a nearby smelter, where attending doctors amputated the shattered leg. He died a few hours later. His remains were buried in the Animas City Cemetery, where they remain.

Incidentally, the physicians who were summoned to aid Stockton submitted a $300 bill to the La Plata County Board of County Commissioners for their services. There is no record of payment. [8, 10, 13, 17, 18, 37, 75, 85, 190, 204, 231, 240, 254]

Stockton, Port (1854–1881). Best known in the San Juan Basin as the younger brother of Ike Stockton, the basin's most notorious gunfighter, Port was a villain in his own right. Indeed, in the annals of the West he is reputed to be a genuine "gunfighter" while Ike gets only cursory mention (see O'Neal's *Encyclopedia of Western Gunfighters*).

Born William Porter Stockton into a family of northeastern Texas ranchers, he was charged with attempted murder as an adolescent. After circulating through Dodge City, Kansas, he migrated to Cimarron, New Mexico, and Trinidad, Colorado, where he got into numerous shooting scrapes and went to jail for the murders of Juan Gonzales and Antonio Archbie. In the first case, he got off on a plea of self-defense. In the second, as was his older brother's practice, Ike came to Port's defense and broke him out of jail.

Riding with a posse in the San Juan Basin, Port shot down a sheepherder with no motive except the knowledge that his companions' dislike of sheep would let him get away with it. As the member of another posse, he cut off the fingers of a just-killed horse thief so he could easily steal the victim's rings. While marshal of Animas City he expressed his dislike for being razor-nicked by shooting at the barber or, according to another version, pistol-whipping him unmercifully. Threatened with arrest for that incident, he left town and headed back to New Mexico. His wife, Irma, and two daughters followed.

On the south side of the Animas River near present Flora Vista, Port stole the claim of recently widowed Mrs. Tom Cooper. Because of such conduct, an underlying dislike for Port simmered, but the final straw came as much from the activities of his henchmen as from his own abhorrent behavior.

To Francis Hamlet's 1881 Christmas party four miles up the Animas River from Farmington came, uninvited, Port's friends Dyson Eskridge, James Garrett, and Oscar Pruett. Outside Hamlet's cabin the intruders shot off their pistols to create a row. When Hamlet and George Brown went to the doorway looking for the source of the commotion, a pistol shot hit Brown and he died on the spot. Return fire from the cabin killed Pruett.

Farmington vigilantes rode up the river to Port's place a few days later, looking for Eskridge and Garrett. They spotted Port at his cabin door and riddled him with a fusillade. Irma ineffectively returned fire. Like the pioneer whose claim Port had jumped, she too became a widow. In revenge for Port's death, Ike killed Aaron Barker, one of Port's assailants. This aggravated a feud between the Farmington vigilantes and Ike, setting the stage for the basin's most famous gunfight. [1, 3, 37, 84, 85, 190, 231]

Stoner, Dolores County, Colorado. A defunct ski area eleven miles northeast of Dolores on Colorado 145.

Stony Pass. This swale on the Continental Divide gave passage to the main route into Baker's Park during the region's 1870s mining heyday. Until the Denver and Rio Grand Railway came up the Animas Valley to Silverton in 1882, mining equipment and merchandise from the east moved on a wagon tollroad up the Rio Grande to its headwaters, then over Stony Pass into Cunningham Gulch, and down to Howardsville. After the railroad's arrival, the pass served primarily as a route for prospectors. Some reports put John Charles Fremont's disastrous fourth expedition at this location during the winter of 1848–1849. Assisted by a team of horses, the first automobile traversed the pass in 1910. Stony Pass (elev. 12,588 feet) is five miles southeast of Howardsville by way of San Juan 4A and 3. [172, 219]

subalpine fir. Like its frequent companion, the Engelmann spruce, this most widespread of the western true firs adorns the basin's mountain peaks from 8,000 to 12,000 feet elevation. When its lower branches dip to the ground under heavy snow, they may take root and form new shoots. Also called the alpine or Rocky Mountain fir, it keeps a pencil-slim form even as it grows uncrowded. In its lower reaches it can grow seventy feet tall; near timberline it may struggle to reach a height of four feet. This fir imitates the blue spruce each spring as its tips display a touch of silver.

Deer and other ungulates browse its bark, songbirds and chipmunks relish its cone-dwelling seeds. Fortunately for them, lumbermen have little use for its soft, coarse wood. [81, 233]

Sullivan, James J. (ca. 1840–?). The La Plata County deputy sheriff who shot and killed Ike Stockton. As Ike's acquaintance, Sullivan had been willing to put up with the outlaw so long as he caused no disturbances within the deputy's jurisdiction and handled his own quarrels. But when Stockton betrayed one of his own gang and broke the Code of the West, Sullivan stopped looking the other way. He used a New Mexico warrant as his authority to get Stockton. And he had the experience and poise to carry out the arrest.

Seeking to retrieve Hi Barber from the protection of the army encampment at Animas City in 1879, he had stood up to garrison commander General George P. Buell. (Barber had shot from his horse Sullivan's friend, Pitt West, and Sullivan demanded that Barber be turned over to civil authorities.) Sullivan's effort was unsuccessful, but his loyalty and cool approach gained Buell's respect.

His demeanor was such that he seldom used a firearm to quell a disturbance. As a policeman in Durango, he had patrolled the high rolling section and marched rough miscreants to jail without touching his weapon. In Rico, where he served as marshal in 1881, Sullivan kept the peace among drunken miners with conflicting claims. There he got wind of a masked gang hiding near town; he suspected they were poised to raid the mining camp's saloons. He preempted their intentions, rounded them up, and assigned one of the local characters to hold the bunch under guard until daybreak.

With this background, he stepped onto the boardwalk of Railroad Avenue on September 26, 1881, confident of his ability to handle Ike Stockton. [51, 186]

Summit Reservoir State Wildlife Area. A private lake built in 1904 by the Summit Irrigation and Reservoir Company at the top of the divide on the Mancos-Dolores Road is this area's main attraction. Owned and operated by shareholders, the 402-acre project is a rare example of pioneer ingenuity without government subsidy.

At 7,388 feet elevation, this facility is situated among willows and ponderosa pines. Water comes to the reservoir from Lost Canyon Creek by way of an irrigation diversion. Its coolwater fishery includes smallmouth bass, rainbow trout, channel catfish, and black crappie. Summit Reservoir is nine miles northwest of Mancos on Colorado 184. [50, 250]

Tabor, Horace A. W. (1831–1899). The silver magnate who got a divorce of questionable legality in Durango so he could marry Elizabeth McCourt Doe, better known as Baby Doe.

Originally a stonecutter from Vermont, Tabor gave up farming in Kansas to join the Colorado gold rush of 1859. As a storekeeper in Leadville, he grubstaked George Hook and August Rieche with sixty dollars' worth of supplies. When they stuck it rich, Tabor got a third of the Little Pittsburgh Mine's $1.5 million profit over the next two years.

Tabor parlayed his share and became a millionaire, civic leader, politician, philanthropist, and philanderer. He underwrote fire departments and public utilities, gave to charities, and built opera houses bearing his name in Leadville and Denver. He was appointed to the U.S. Senate and served as Colorado's lieutenant governor.

At the height of his wealth, he fell for the twenty-six-year-old divorced Baby Doe and secretly sought a separation from his first wife, Augusta, in remote Durango. (As the owner of the H. A. W. Tabor Pioneer Stage and Express Line out of Durango and with mining interests in the San Juans, he easily camouflaged his real reason for being there.) Before the divorce was final, he wed Baby Doe in St. Louis.

When the value of silver plunged in 1893, many of Tabor's mines, which served as collateral for a web of other investments, became worthless. Even his favorite mine, the Matchless, fell to creditors. His financial house of cards collapsed. Unpaid utilities stopped service at Tabor's palatial Denver residence; Baby Doe sold her jewelry to put food on the table.

Destitute, Tabor worked as a common laborer in Leadville, then used political connections to become Denver's postmaster. Myth has it that on his deathbed he remembered his favorite mine and told Baby Doe,

"Don't give up the Matchless." Whether he whispered a word about his former holdings in the San Juan Basin is unknown. [111, 113]

Tacoma, La Plata County, Colorado. At this defunct community on the Animas River, the La Plata Electric Association's Tacoma hydroelectric plant generates power from water that flows down the mountainside through a pipe from Electra Lake. The waters of Cascade Creek, diverted through a 4,000-foot flume, feed the lake. Built in 1905–1906 by the Animas Canal, Reservoir, Water Power Investment Company to serve a population that had not yet come, the original plant soon went into receivership.

Before the Depression, the Tacoma community was home to about seventy-five people. Tacoma is twenty-two miles north of Durango near milepost 472 on the Durango and Silverton Narrow Gauge Railroad. [11, 95, 101]

Taos County, New Mexico. Now confined to the north-central part of the state, this county once stretched through the San Juan Basin and all the way across New Mexico Territory, whose western border then touched California. After Congress split New Mexico to form Arizona Territory, part of Taos County joined Rio Arriba County. Following numerous county boundary realignments, no portion of Taos County remains in the basin. [247]

territory. A political unit invented to secure the federal government's takeover of the San Juan Basin and other parts of the West. Three such jurisdictions have governed the basin. Congress was well aware that mistreated colonies might rebel and separate—some of its members had signed the Declaration of Independence in defiance of the British crown. To control its own western American colonies and avert potential disobedience, Congress set forth a scheme in the Northwest Ordinance of 1787 that balanced liberty with imperial expansion: the American territorial system.

Through progressive stages, territories graduated from management by appointees to government by elected representatives. In its advanced form, the internal workings of a territory were much like those of a state. Statehood—and seats in Congress—followed at Congress's option.

After the United States got control of the San Juan Basin through the Mexican Cession of 1848, Congress divided the ceded region between Utah Territory in the north and New Mexico Territory in the south by running a line along the 37th Parallel—across the middle the basin. Eleven years later Congress cut away the part of the basin lying within Utah Territory to include it in the newly formed Colorado Territory.

Thus, citizens of the basin have lived under New Mexico, Utah, and Colorado territorial governments. [78]

Thornburgh Battle. See Meeker Massacre.

Tierra Amarilla Land Grant. The most famous and controversial of the Mexican land grants. Located partly within the San Juan Basin, it illustrates the deceit and skulduggery associated with Spanish and Mexican methods of transferring real estate from public to private ownership—the land grant system.

After separating from Old Spain in 1821 and forming the Republic of Mexico in 1824, Mexican rulers carried on the practice of giving land to their settlers as a means of discouraging intrusion by Native Americans and U.S. citizens. By the 1848 Treaty of Guadalupe Hidalgo, which concluded the Mexican-American War, the United States took possession of the region and agreed to respect the property rights of those who had been Mexican citizens. Congress delayed setting up the Court of Private Land Claims to sort out conflicting requests until 1891. To pursue land title in the forty-three-year interim, claimants petitioned the Surveyor General of New Mexico and sought an act of Congress.

Having settled in Rio Arriba country, one Manuel Martínez petitioned New Mexico governor Santiago Abreu for such a grant in 1832. After the United States took over the territory, his grant documents arrived at the desk of Surveyor General William Pelham. Among the questions Pelham had to decide was whether the grant was to a community, with a common area, or to Martínez as an individual. While acting as lawyer, judge, and jury, he relied on inept or dishonest staff reports, ignored due process, discarded inconvenient evidence, and decided this was an individual transfer. Congress subsequently confirmed a 929-square-mile tract to Martínez's son, Francisco. Francisco Martínez deeded parcels to other settlers.

Subsequently, Thomas B. Catron bought titles from Martínez, knowing that many papers of ownership conflicted with other deeds but confident that he could manipulate through the legal muddle and acquire ownership. Through such connivance he became the largest landowner in the nation. In this case, as in several hundred others, historical puzzles and precarious titles opened the door for legal shenanigans that persist to this day. [44, 129]

Tiffany, La Plata County, Colorado. This was a small farming community in the 1900s. Tiffany (elev. 6,200 feet) is three miles west of Allison on Colorado 151.

Tison Gang. A band of outlaws that spread fear through the San Juan Basin and the Southwest in 1978. On July 30 of that year, with the help of his three sons, Gary Tison walked out of the Florence, Arizona, prison where he was serving two consecutive life sentences for killing a prison guard. He, his sons, and fellow escapee Randy Greenawalt then rampaged through New Mexico, Colorado, and Arizona. The five abducted a Texas couple, James and Margene Judge, at South Fork, Colorado, triggering a massive manhunt. They executed the couple off a back road south of Chimney Rock.

After camping near Dolores, the group headed back to Arizona where their exploits ended in a barrage of bullets short of the Mexican border. One of the Tison boys was killed. The other two and Randy Greenawalt were captured. Tison escaped in the melee, but his body was found a few days later in the desert where he had died a tortured death, broiled by the Arizona sun. The degree of guilt of the sons, who did as their domineering father told them, remains a subject for debate by lawyers and psychologists. [34]

Toadlena, San Juan County, New Mexico. Bob Smith and his brother, Merritt, built this trading post in 1909. Toadlena and its trading post are eleven miles west of Newcomb on Navajo 192. [87]

Tocito, San Juan County, New Mexico. Trader Jess Foutz built a trading post here in 1913. The post is nine miles north of Newcomb on U.S. 666, then three miles west on Navajo 56. [87]

Totten Reservoir State Wildlife Area. Located east of Cortez in the McElmo Creek drainage, this agricultural impoundment covers 249 acres at an elevation of 6,158 feet. Its warmwater fishery includes bluegill, largemouth bass, walleye, northern pike, yellow perch, crappie, sunfish, and channel catfish. Administered by the Colorado Division of Wildlife, Totten Reservoir is three miles east of Cortez on U.S. 160 and a mile north on Montezuma 29. [250]

Towaoc, Montezuma County, Colorado. The Ute Mountain Ute Tribe's headquarters and the Bureau of Indian Affairs' Ute Mountain Ute Agency are housed at this community. A two-mile-long road to Towaoc (1990 pop. 700, elev. 5,880 feet) goes west off U.S. 160 eleven miles south of Cortez.

trading post. A place where white traders exchanged goods for Native American products. Although the term applies to a place of trade with

Emmet Wirt Trading Post, ca. 1895. When a land shark filed on land where trader Emmet Wirt and others had squatted, the settlers avoided a showdown by moving to this location, which became Lumberton. Courtesy La Plata County Historical Society.

the members of any tribe, most in the San Juan Basin's posts were associated with Navajos.

After the Navajos were permitted to leave Bosque Redondo in 1868 and return to land set aside as a reservation, comprising about one-tenth of the land they claimed, white merchants entered the area to sell them food and other provisions. Although the use of coins as a medium of exchange was not unfamiliar to them, the Navajos had little or no money, so barter served as the basis for the economy. The traders accepted wool, livestock, and sand paintings in exchange for merchandise they freighted to the posts. From this system the Navajos' modern rug business matured.

Most trading posts were far from any town. Like the saloonkeeper of the mining areas, the trader assumed many roles. Since few Navajos learned English, traders learned to speak Navajo and, as translators, acted as intermediaries between Navajos and government or business agents.

A string of posts ran south from Shiprock seventy miles along the east edge of the Chuska Mountains: Sanostee, Toadlena, Two Gray Hills, Newcomb, Sheep Springs, Naschitti, Tohatchi, Mexican Springs. Many continue today, some as trading posts, others as general stores. Although some business may still be transacted by barter, the Navajos have developed markets other than trading posts for their products and sell most of their goods for cash. Nevertheless, the term "trading post" persists, used in some instances as a marketing gimmick to give to the tourist an illusion of authenticity. [82, 93, 106]

Trail of the Ancients. This route extends from Four Corners through Cortez, Dolores, and Pleasant View and then heads west for Utah. The Colorado Scenic and Historic Byways Commission has designated the route a byway. [209]

trance rite. Some Native American tribes rely on members with psychic powers to sort out good and evil or get rid of bad spirits. Among the Navajos lived Hosteen Beaal, who could bring forth information from a trance. This power was demonstrated when he helped white men solve a crime committed by some of his own people.

At Jack Jones's trading post, two potential customers found the place ransacked with the merchant's body wrapped in a blanket under the store counter and his head under the kitchen stove. A meat cleaver lay on the kitchen floor. The investigating sheriff surmised what had happened, but he had no way of telling who the killers might be. He offered $500 as a reward for information leading to the criminals' capture.

Franc Newcomb talked the sheriff into commissioning Beaal to find out who did it. Beaal held a trance rite and told this story after he woke up.

Two Navajos bought liquor from a bootlegger in Gallup and rode out of town. By the time they reached Jones's trading post, they had consumed the liquor and wanted more. The merchant realized they were drunk and dangerous, so he told them he was sold out. The Navajos started searching the place anyway. When one of them saw Jones reaching for a gun, he grabbed a meat cleaver and swung such a powerful blow that he severed the trader's head and sent it flying through an open door into the kitchen. After searching the place for liquor and taking what they wanted, the Navajos carelessly hid the body, rolled the head under the stove, and rode off to visit their relatives in the Chinle Valley. That's where the sheriff could find them, Beaal said.

Beaal even went further. Although he could not mention names, he described a scar on one of the individual's cheeks and said the other was short with a lame hip. Within four days, the sheriff had the culprits in custody.

Beaal received a check for $100. Who got the rest of the reward is not of record. [93]

trapper. To harvest fur-bearing animals, the mountain men of the early nineteenth century used the east-flowing tributaries of the Mississippi River—the Missouri, Yellowstone, Platte, and Arkansas Rivers—to fan out west from St. Louis. The San Juan Basin was, therefore, on the southern edge of the mountain men's world, a latecomer to the fur trapper's domain. To reach the basin, the trappers were faced with the task of rowing up the Arkansas, climbing over the Sangre de Cristo Mountains, crossing the Rio Grande valley, and surmounting the Continental Divide. Others took an overland route that became the Santa Fe Trail, then trapped north and west from Santa Fe to reach the basin.

They would take minks, martens, otters, and foxes, but all of these together lacked the attraction of the beaver. Among the trappers who scoured the basin were Ewing Young and William Wolfskill. When Young came over the Continental Divide to trap along the San Juan River in 1824–1825, he likely followed the southern trace through Santa Fe. Wolfskill, who preferred the Dolores River, also came by way of New Mexico.

The beaver habitat of the San Juan Basin presented transportation problems unfamiliar to the former Mississippi Basin trappers. Many rivers, standard highways for the mountain men, were too treacherous for navigation. The lack of usable waterways compelled the basin fur traders to use pack animals. The southernmost of the well-known fur rendezvous was at Henry's Fork on the present-day Utah-Wyoming border.

Some of the basin's trappers may have gathered there to sell their annual harvest of 300 or 400 pelts. More likely, they took their goods into New Mexico.

If they did, instead of joining other mountain men at a rendezvous, they missed a soiree of dauntless proportions. The trappers, traders, buyers, and hangers-on got together yearly, normally after the spring hunt, to participate in a Rocky Mountain version of the medieval fair. Using beaver furs as their medium of exchange, the trappers grabbed up supplies for the coming year—gunpowder, traps, blankets, lead, tobacco, foodstuffs, trinkets for concubines. Any skins left over went for whisky, which magnified an eruption of fighting, roaring, racing, stealing, gambling, feasting, and cheating, not to mention debauchery with native women.

Trapping the beaver, whose fur went mostly for men's hats, reached its height in the early 1830s. By the mid-1840s, a combination of trade that brought substitute materials, the depletion of beaver, and a change of fashion from fur to silk ended the era of the mountain man and western fur trade. [35, 52]

tree species. Between its low dry desert and moist alpine peaks, the basin's five vegetative zones nourish a cross section of trees that grow in the Mountain West. Interrupted by narrowleaf cottonwoods along stream banks, piñon and juniper scatter across the lower arid regions, often intermingling with Gambel oak; you rarely find them above 8,000 feet elevation. At about 6,500 feet the ponderosa pine and quaking aspen spread across shaded, moist slopes. A little higher up you may view Douglas fir, then blue spruce through their range to 11,000 feet. Beginning at about 8,000 feet the Engelmann spruce and subalpine fir mix with the other evergreens and stretch to their 12,000-foot tree lines. [81, 233]

Trimble Springs. A favorite spa for Native Americans long before William Frank Trimble settled there in 1874 to treat his Civil War wounds in its 110-degree water. Pleased with the result, Trimble built a boardinghouse and way station. The springs soon became known for their healing powers. Acquired in 1882 by Tom D. Burns, a New Mexico freighter, entrepreneur, and politician, the place became a resort as well as a health spa. The Burns family (founders of Durango's Burns National Bank) operated or leased the place until 1931. It reached its heyday around 1900. For more than a century, discounting interruptions by fires (1892, 1931, 1957) and wars, the spa has been a favorite resort for residents and travelers alike. In the 1980s the resort was again rejuvenated and today features an Olympic swimming pool.

Trimble Springs is nine miles north of Durango on La Plata 203, just off U.S. 550 via La Plata 252. [27]

tumbleweed. Although romanticized in cowboy songs as a "tumbling tumbleweed," this weed is a nuisance to farmers and stockmen and a threat to more useful vegetation. In the fall the plant's stems break off near the ground so the wind can roll and tumble it across the prairie to spread seeds. The Russian thistle, a native of Asia, is a common tumbleweed in the San Juan country. [38]

Turley, San Juan County, New Mexico. On the northwest bank of the San Juan River, in Largo Canyon upstream from present-day Farmington, Hispanic settlers started a nineteenth-century community they called Alcatraz. When it was moved to the other side of the river, it got the name of its postmaster, Turley. Neither community remains. [3]

Twin Buttes. These two buttes rise to the north of the banks of the San Juan River between Kirtland and Farmington. Viewed from the south, they reveal why the Navajos call the formation *Asan bi pey,* "Where-the-milk-comes-from-a-woman." [85]

Twin Crossings. The Animas River cuts a U-shaped bend to the west and into a steep bank at this place. As a result, pioneers traveling the original wagon road on the west side of the river had to make twin crossings—ford the stream to get to the east (opposite) bank, and then cross again to get back on the west side. This location served as a staging area for the white settlers who entered the Ute Strip in 1899 to stake land claims.
    Twin Crossings is on U.S. 550 twenty-one miles south of Durango. [191]

Two Gray Hills. Situated on a barren plain east of the Chuska Mountains, this post is known for its solitude and the fabrics woven by the Navajos who live there. Using only black dye, the artists blend natural white, beige, tan, and gray wool tones to weave the most prized, and expensive, of all Navajo rugs. First opened in 1897, Two Gray Hills (elev. 5,920 feet) is five miles west of Newcomb on Navajo 192. [71]

Uncompahgre. A Ute band, also known as the Tabeguache, that hunted primarily in west-central Colorado (not within the basin) but was removed to Utah's Uintah and Ouray Reservation in 1880. The band is relevant to the basin because its chief, Ouray, became leader of all Utes. [86, 119]

unidentified flying objects (UFOs). Of the 12,618 UFO reports received by the air force from 1947 to 1969, perhaps none was more dramatic than the sightings at Farmington in March 1950. According to newspaper reports, "fully half" of the townspeople were certain that they saw strange space ships. Some observers reported a few, others estimated 500 were in the flock. Harold E. Thatcher, head of the local Soil Conservation Office and an amateur engineer, did some trigonometry and estimated the alien aircraft whizzed 20,000 feet over the basin at not less than a thousand miles per hour. No one has proved the existence of a UFO, over the San Juan Basin or elsewhere. [15, 217]

Upper Colorado River Basin Compact. An agreement reached in 1948 between the states of Colorado, Wyoming, Utah, Arizona, and New Mexico that allocates Colorado River water among those states. It was a necessary offspring of the Colorado River Compact, which allocated water between Upper Basin and Lower Basin states. [263]

uranium. The by-product of vanadium that is used to make atomic energy—and bombs; it has periodically spurred the economy of the basin. In 1941 the Defense Plant Corporation, a World War II agency, financed the construction of a vanadium mill at Durango that operated until the war ended in 1945. Under threat of the Cold War, the Atomic Energy Commission reactivated the mill in 1949 to make uranium concentrate

Utes, ca. 1920. The ancestors of these Southern Utes were members of the Capote and Mouache bands that hunted southern Colorado and raided the camps of their traditional enemies, the Navajos. Courtesy Fort Lewis College, Center of Southwest Studies.

from southeastern Utah's ore collection stations. And to make sure the uranium got safely to Durango's mill, the commission built over 200 miles of roads.

The Vanadium Corporation of America acquired the mill in 1953, then moved its operation to Shiprock in 1962. As the government propped up the basin's urban economy more than half a century before, when it bought silver, it now repeated the favor as the sole consumer of uranium. [55, 113]

Ute. A tribe of loosely confederated Native American bands whose original domain encompassed most of Colorado and Utah and spilled into New Mexico and Wyoming—a 150,000-square-mile mountain wilderness. Most historians place the number of bands at seven: the Mouache east of the San Juan Basin; the Capote in the San Louis Valley and Rio

Arriba; the Weminuche in the basin; the Uncompahgre (Tabeguache) north of the basin; the Grand River (Parianuc) along the Colorado-Utah border; the Yampa in the Yampa River valley; and the Uintah in Utah's Uintah Valley.

By custom, each of the bands used a defined hunting territory that it defended against non-Ute tribes. The Utes have lived continuously in Colorado, and perhaps in the New Mexico portion of the San Juan Basin as well, longer than any other residents. Today the United States recognizes three Ute tribes: the Uintah in Utah and the Southern Ute and Ute Mountain Ute in the San Juan Basin.

Their ancestors came to the basin and the rest of their domain after moving south along the East Slope of the Rocky Mountains. Perhaps they moved to their mountainous territory to avoid other native tribes: the Apache, Arapaho, Cheyenne, Comanche, Kiowa, Pawnee, and Sioux to the east and northeast; the Apache (except the Jicarilla Apache, who were friendly) and Navajo to the south; the Bannock, Goshute, Shoshone, Snake, and Paiute to the west and northwest. The Ute, the only tribe considered native to Colorado, remained isolated from other Native Americans by preference and terrain.

The Spanish were the first Europeans to affect the Utes; they brought horses in the 1600s. Once on horseback, the Utes found it easier to raid other tribes' livestock than to hunt wild game. As they became raiders and more warlike, the Utes spread out from their mountain surroundings. Even so, they managed to remain divorced from other tribes. And until the 1860s the Utes avoided Spanish and Anglo intruders as well, although they did ally with the Spanish against the Navajo, Comanche, and Apache tribes.

Then came the gold seekers of the 1870s. Even after the white's initial invasion, the Utes were able to sustain a semblance of their way of life and avoid reservation confinement longer than most tribes. Through Chief Ouray's negotiating skills the tribe kept possession of nearly 1,900 square miles through the 1870s by letting whites occupy the mineral-rich portion of the San Juan Mountains. This agreement, signed by President Ulysses S. Grant in 1874, was called the Brunot Agreement, after Indian Commissioner Felix Brunot, or the San Juan Cession. (It wasn't a treaty because Congress had stopped making treaties with Native Americans.) It blew apart only five years later, in 1879, when disgruntled Utes at the White River Ute Agency in northeast Colorado slaughtered eleven Anglo men and ran off with five females in what is loosely called the "Meeker Massacre." The ensuing white panic prompted military action.

The United States ordered the four northern Ute bands to gather near Grand Junction, Colorado; from here the army marched them to the largely barren Uintah Reservation in Utah, far removed from their beloved shining mountains. The Mouache, Capote, and Weminuche bands were herded into southwestern Colorado reservations. Chief Ouray and his comrades could delay no longer; the noble Utes at last joined their fellow Native Americans in a state of final subjugation.

Today the Utes have three reservations, the Southern Ute and Ute Mountain Ute in the basin and the Uintah in eastern Utah. [74, 86, 119]

Ute Mountain Tribal Park. This facility features ancestral Pueblo ruins similar to those in Mesa Verde National Park but is more than three times as large. Located on tribal lands to the south and west of the national park and operated by the Ute Mountain Utes, it offers guided tours to the public. The park's visitor center is twenty-two miles south of Cortez on U.S. 666. [165]

Ute Mountain Ute Reservation. A 922-square-mile territory where, by an 1895 agreement, the Weminuche band of Utes retained land in common for their exclusive use. It contains both tribal trust lands and tribal fee properties. Congress split off the Weminuche's portion to form the Ute Mountain Ute Reservation. It stretches forty miles along the Colorado–New Mexico border from Utah (and the Navajo Reservation) to the west edge of the present Southern Ute Reservation, a point roughly south of Mancos. All except a 200-square-mile New Mexico portion lies in Colorado. The original tract was rectangular. In 1911 the band (recognized by the United States as the Ute Mountain Ute Tribe) ceded land for Mesa Verde National Park in exchange for other lands. This exchange produced the reservation's irregular boundaries. [42, 74, 119, 220]

Ute Mountain Ute Tribe. An organization of Utes that controls the Ute Mountain Ute Reservation, where 1,500 of its members live. The tribe is headquartered at Towaoc. About 250 tribal members live near Blanding, Utah. Ute Mountain Utes are descendants of the Weminuche band that hunted throughout the San Juan Basin portion of the Utes' native territory. The band became a separate tribe primarily because, unlike the Mouache and Capote bands, it opposed the allotment of land from the United States to individual members. In protest, the Weminuche band set up its own enclave within the Southern Ute Reservation. [74, 119, 220]

Ute Pass. This crossing in La Plata County, Colorado, should not be confused with eight other passes of the same name within the state. Dividing the Spring Creek and Florida River watersheds, it has been used by natives, military expeditions, trappers, prospectors, loggers, and tourists. Ute Pass (elev. 7,420 feet) is on La Plata 240 (known locally as Florida Road) eight miles east of Durango. [219]

U

Ute Strip. A fifteen-mile-wide section of land running along the north side of the Colorado–New Mexico border from the Continental Divide westward to Utah. At high noon on May 4, 1899, the land office opened the eastern portion of the strip to white settlement. Allison, Bondad, Ignacio, Oxford, Tiffany, and part of the Florida Mesa lie within the sector. On most maps the area is broad-brushed as part of the Southern Ute Reservation. What is behind this apparent contradiction?

While the Weminuche band preferred tribal, in-common ownership and were granted lands farther west, members of the Mouache and Capote bands received family allotments. Unalloted land was thrown open to white settlement. That is why today the strip portion of the Southern Ute Reservation is checkered with non-Ute-owned land.

Even though the opening of the strip brought national attention, the land office failed to make clear the conditions for taking the available land. Some whites thought they were supposed to make a run north from the New Mexico state line in the manner of the Oklahoma land rush. Others wanted to occupy a section and preempt the land. Still others filed claims at the Durango office. Those who charged into the strip to drive down a stake found themselves in conflict with others who had the same idea. After the confusion was cleared up, most of the land seekers were disappointed. Ute families had already received the best agricultural parcels. [74, 174, 177, 187, 191]

Vallecito Reservoir. See **Pine River Project.**

vegetation zones. If you go from the San Juan Basin's lowest elevation at about 5,000 feet up to one of its 14,000-foot peaks, you gain 9,000 feet. In climatic terms, gaining 1,000 feet is like going 600 miles northward. By taking advantage of the elevation differences, you can go on a 5,400-mile trip without leaving the basin; that's like going from New Orleans to Fairbanks.

Fanciful analogies aside, the basin's vegetation lives in as many zones as you are likely to find anywhere within a region of similar size. But in addition to elevation, soil conditions, precipitation, and the slope of the land determine zone boundaries. Fingers of zones extend into one another and overlap, so you can't identify a zone by its elevation alone. On the way up a mountain, a zone begins on a south slope several hundred feet higher than on terrain slanted north. You must look at the plant community. (Keep in mind also that biologists use different names and elevations among themselves to draw zone boundaries.) Water diversions and other human activities can also change zone locations.

The Upper Sonoran Zone begins at about 4,500 feet and prevails in the basin's lowest elevations—the New Mexico portion. Where there is little rainfall or the soil fails to hold moisture, the conditions support little but sage and other arid-tolerating brush. Piñon and juniper with a rough carpet of blue grama and buffalo grass live in the zone's upper foothills and valleys.

Where the flatlands break in transition to the large trees of the mountains, at altitudes from 5,700 to 8,000 feet, the Foothills or Transition Zone takes over to add Gambel oak, kinnikinnik, and other woody shrubs to the piñon and juniper. Generally, the higher the elevation within this zone, the more scrub oak and the less piñon-juniper you find.

The Montane Zone begins at about 8,000 feet, forested with ponderosa pine on its dry south-facing slopes and Douglas fir on the moist north-facing slopes. Hardy grasses and wildflowers flourish beneath aspen in this zone.

The Montane Zone merges into the Subalpine Zone at altitudes between 11,000 and 11,500 feet. Here Engelmann spruce, Douglas fir, and subalpine fir shelter what some describe as the flower garden of the Rocky Mountains. This is the area most enjoyed by the basin's outdoor enthusiasts. When you view miles of spruce-fir forest covering the mountainside, perhaps with accents of intruding aspen groves along the lower water courses, you are looking at subalpine forest. The forest floor is soft with rotting needles and decaying tree trunks. The snow that has accumulated all winter is sheltered and stays late for the dainty, moisture-loving flowering plants that flourish in the cool shade of the evergreens. At the timberline region of the zone, where the trees are less dominant and the snowdrifts melt slowly, wildflowers cover the slopes in profusion.

Above timberline begins the region of rock fields, or scree, and grasslands. This is the Alpine or Tundra Zone. The snow melts from its exposed slopes more rapidly than on the shaded areas below, so spring comes earlier in this higher region. Indeed, some of the zone is swept clean of snow during the winter and offers little moisture for its sparse plant life. But the melting edges of perpetual snowdrifts, trapped in draws and depressions, nurture flowers throughout the short summer. The predominant vegetation of the mountain tundra forms a tight, short turf that, when well established, excludes other plants. But rock outcrops and gravel slopes provide ample space for flowering plants to form their natural gardens. Somewhere between 12,000 and 14,000 feet the flowering plants give up, and only lichens survive. [28, 207, 225]

vigilantes. Self-proclaimed law enforcement committees—like the bunch that gunned down Port Stockton at his place on the Southside Road between Aztec and Farmington in 1881. Such gangs were ostensibly organized to protect life and property against outlaws and punish wrongdoers in the absence of official law enforcement. Although the tradition of citizen vigilance goes back to colonial times and is commonly associated with the Forty-Niners' gold rush, the practice was widespread throughout the West and the basin.

The Farmington Stockmen's Protective Association, whose members murdered Stockton, was formed to combat rustling. The group's initial motives were honorable. After all, the place was still part of Rio Arriba County and it could take a week to get the sheriff from the county seat at

V

Tierra Amarilla. But without legal restraints the group became a device for personal revenge.

Durango's Committee of Safety protected the community from New Mexican "invaders" as well as local miscreants. Among its most notable deeds (or misdeeds) was the lynching of Henry R. Moorman for shooting a gambler. Its spurious claim as a force for good was laid bare, however, when its 300 members used their bandanas as masks. [3, 49, 130]

wapiti (elk). Why should the basin's largest deer be called a wapiti instead of an elk? Because people elsewhere in the world use "elk" to refer to the animal we call a moose. Before European settlers arrived, wapiti were plentiful in all areas of the Southwest with suitable habitat, including the plains. Market hunting drove them nearly to extinction in the early twentieth century. Transplants from Yellowstone replenished some of the herds. Wapiti now range in the basin's mountains and high meadows where they like to graze in herds—sometimes in groups of several hundred. [135]

Washington Pass. In 1835 2,000 New Mexicans left Santa Fe and headed west through this narrow gorge in search of Navajo slaves. Lining the trail behind rocks and bushes, 200 Navajos waited. At the signal of an owl's hoot, the warriors let go a rain of arrows. Few New Mexicans lived to straggle back to Santa Fe. In 1849 Colonel James Macrae Washington maneuvered his troops through here and described it as the most formidable defile he had ever seen. (Some recent maps label the pass "Narbona" after the Navajo warrior.) Simpson reported that its most narrow passageway was only fifty feet wide with a bluff rising 600 feet on the north. Colonel Kit Carson went through the canyon in his efforts to subjugate the Navajos.

Washington Pass (elev. 9,365 feet) is six miles east of Crystal (midway between Crystal and Sheep Springs) on New Mexico 134. [82, 204, 243]

Waterflow, San Juan County, New Mexico. On a mesa near this place three Catholics (so identified because they were not among the preponderance of Mormons settling the area) built and shared a sandstone dugout in 1909. Next to a low ridge of bluffs in the valley below lay a community called Jewett. For some obscure reason the location is now called Waterflow.

Waterflow (1990 pop. 300) is ten miles west of Farmington on U.S. 64. [85]

Watermelon Bust. A forerunner of the San Juan County, New Mexico, Fair. The pioneers got together for an outing each fall on the Animas River east of Farmington. When agriculture matured, farmers displayed fruit and other produce. The picnic became a fair in the 1880s. [3, 192]

weed species. You can get in trouble with the law if you let certain non-native plants grow on your property in Colorado, and they're not welcome in New Mexico either. They steal moisture, nutrients, and sunlight from surrounding native plants. The leafy spurge, for example, has roots that go down thirty feet. A perennial, native to the Ukraine, it came to North America in 1827 and has been a plague ever since. It is so aggressive that it can crowd out all other vegetation. Not only that, its milky latex can damage your eyes and skin. It is one of four weeds that, by law, a landowner in Colorado cannot ignore. The others are three knapweeds: the diffuse, the Russian, and the spotted. Violators don't go to jail for not controlling the weeds, but the law emphasizes the weed problem and backs up local weed control authorities. Other plants on the weed controller's most wanted list are the purple loosestrife, yellow and Dalmatian toadflax, Canada thistle, poverty sumpweed, water hemlock, whitetop, whorled milkweed, hound's tongue, and musk thistle.

While some native plants are troublesome (the larkspur is poisonous to cattle), the most damaging weeds came from Eurasia. Of Colorado's thirty worst weeds, half migrated from that part of the world. [38]

Weminuche. A Ute band whose hunting territory included the western San Juan Basin and a region north of Four Corners. Many of their descendants are now members of the Ute Mountain Ute Tribe. [119]

Weminuche Wilderness. With over 700 square miles, this is one of the National Wilderness Preservation System's largest areas. Among its labyrinth of rugged mountains, scores of peaks rise over 13,000 feet in elevation. Three peaks go up more than 14,000 feet: Windom Peak, 14,091; Mount Eolus, 14,086; and Sunlight Peak, 14,060. Its rains and snows produce nearly 800,000 acre-feet of water yearly for the Colorado River watershed. Set up in a 1975 law signed by President Gerald Ford and stretching over four counties, the Weminuche Wilderness lies next to the meandering Continental Divide within the San Juan National Forest. [159]

wildflower species. The basin's yucca plant provides more than New Mexico's state flower. Its fibrous, sharp-tipped leaves have earned it the name Spanish dagger. Its fruit feeds a variety of animals, and its rosette shelters several bird species. Native Americans have used its roots to make soap and its fibrous leaves to weave clothing. Colorado's state flower, the Rocky Mountain (blue) columbine, is known only for its beauty. The basin's showiest gardens appear soon after the snow melts above timberline. That's when subalpine varieties of daisies, bluebells, and buttercups burst forth with other mountain flowers in mixed profusion.

But all that blooms is not rosy. The larkspur is poisonous to cattle. The Canada thistle, introduced from Europe, sometimes appears in the foreground of pictures selling real estate. It is a noxious weed. [28, 38, 207, 225]

wildlife. Encompassing five vegetation zones starting at about 5,000 feet elevation and rising to 14,000 feet, the basin offers habitat for a wide variety of wildlife. Perhaps a hundred species of mammals live there, ranging from shrews and chipmunks to mountain lions and black bears. In the fall hunters scour the mountains and foothills for deer, wapiti (elk), and black bears.

The beaver brought Anglos to the basin early in the nineteenth century as trappers took advantage of a fad for beaver-skin hats. Like bobcats and coyotes, beavers are still trapped for their fur. Miners and settlers hunted wapiti and deer for food. By the time hunting regulations took effect in the early 1900s, the wapiti had been all but wiped out. Game officials imported specimens to replenish the herds.

Most dramatic among the basin's several hundred bird species are its eagles and hawks. Bald eagles scout the rivers, golden eagles nest in its rocky cliffs, red-tailed and other hawks glide through the canyons.

In the woods you are likely to see less glamorous species, such as the curious camp robber. More officially known as the gray jay or Canada jay, it waits to steal a tidbit. More glamorous is the Steller's jay, blue with a high crest. The talkative, long-tailed magpie becomes an exercise in contrast when it spreads its green iridescent wings to display white tips. Magpies like piñon-juniper country, especially near streambeds.

In the basin's lower reaches, New Mexico's state bird, the comical roadrunner, dashes about looking for lizards and insects. Colorado's state bird, the lark bunting, is most commonly found on the prairies east of the Continental Divide, but you can also find it on the basin's grasslands. The band-tailed pigeon was a favorite game bird of white pioneers. In some communities a band-tailed feast was a yearly event. Blue grouse, mourning doves, chukars, sage grouse, and wild turkeys also

White-crowned sparrow. This songbird uses the basin's diverse habitat to its advantage. It can winter at lower elevations, then migrate a short distance to breed. It is conspicuous and abundant throughout the West. Courtesy San Juan National Forest.

challenge the sportsman. The menial raven holds a special place for some residents. As in many Native American cultures, this large member of the crow family appears in Navajo legends.

Although much of the basin is semiarid, its streams, lakes, and reservoirs attract waterfowl and shore birds. Brown pelicans visit Morgan Lake.

As for reptiles, basin habitats suit the short-horned lizard; it thrives in the mountains as well as on the arid plains. The sagebrush lizard lives where you think it would. You are more likely to see the nonpoisonous bullsnake than any other snake. Garter snakes slither through gardens and enjoy streambeds in the basin, just like they do in the East.

The prairie rattlesnake is the largest rattler in the basin; it can grow to a length of five feet. Its range extends into the basin from the Great Plains. Two lesser-known rattlesnakes live in the western part of basin. The Hopi rattlesnake and the midget faded rattlesnake seldom grow longer than two feet.

The premium game fish in the basin is the cutthroat trout, the only trout native to the region. The brown, rainbow, brook, and lake trout are imports. Wildlife agencies have also brought in warmwater fish—bass, crappie, perch, catfish—to stock reservoirs.

Before dams stemmed its flow and human activity polluted its waters, the San Juan River provided habitat for monsters called Colorado squawfish. Some specimens grew five feet long and weighed eighty pounds. The squawfish, the humpback chub, the bonytail, and the razorback sucker were once regarded as "trash" fish because they compete with exotic game fish for food and habitat. Fishermen detest them. Wildlife agencies have poisoned them. Now they are on the federal list of endangered species.

wildlife areas. States maintain wildlife areas to preserve wildlife for hunting, fishing, and viewing. Protected wildlife areas give the sportsman advantages over other places. For example, farmers may demand the release of water from irrigation reservoirs though the fish may suffer. Not so if a lake, or at least a conservation pool, is set aside to house fish. (But with the West's convoluted water laws you can always find exceptions.)

Within the basin the Colorado Division of Wildlife and the New Mexico Department of Game and Fish operate several sites: Andrew's Lake, Bodo, Devil Creek, Echo Canyon Reservoir, Fish Creek, Groundhog Reservoir, Haviland Lake, Jackson Lake, Joe Moore Reservoir, Lone Dome, Narraguinnep Reservoir, Navajo, Pastorius Reservoir, Perins Peak, Puett Reservoir, Summit Reservoir, Totten Reservoir, William's Creek Reservoir. There are no national wildlife areas in the basin. [250]

Wilkinson, Bert (ca. 1860–1881). The trusted Stockton-Eskridge gang lieutenant whom Ike Stockton betrayed to collect reward money. His family settled near Farmington, where he was known as a willing helper to Charles A. Jones, his neighbor and later publisher of Rico's *Dolores News*. But signs of lawlessness soon emerged. In January 1879 Wilkinson was arrested for shooting off his gun within the city limits of Animas City and fined five dollars. One Christmas night he killed a character named "Comanche Bill" in Durango.

Bert Wilkinson, ca. 1880. When authorities offered a reward for the arrest of this fugitive after he killed Silverton's town marshal, his outlaw gang leader, Ike Stockton, double-crossed him and collected the money. Courtesy Farmington Museum, 1985.430.1.

At some point he became a partner with Ike Stockton and Harg Eskridge in working a mine. Then he fell in with the Stockton-Eskridge gang and gained a reputation as a reckless daredevil. In August 1881, Wilkinson was in Silverton with Dyson Eskridge and "Black Kid" Thomas. There he was confronted by Town Marshal D. C. "Clate" Ogsbury. (Reports differ as to whether Ogsbury was quelling a row or executing an arrest warrant.) One of the three fired a shot that killed the marshal.

Kid Thomas, having taken no part in the incident, made no effort to get away. The townspeople took him into custody and lynched him the next morning.

Eskridge and Wilkinson were cut off from their horses but escaped into the mountains on foot and hid in the outbuilding of a stage-line changing station somewhere between Silverton and Rico. Incensed at the murder of their well-liked and respected marshal, the Silverton townspeople offered a substantial reward for the killer (reports vary from $2,000 to $4,000.) Based on his reputation and witness reports, the search centered on Wilkinson.

Upon learning of the reward, Stockton discovered Wilkinson's whereabouts (not difficult in the case of a fellow gang member) and arranged to be alone with him. Stockton then took Wilkinson into custody and delivered him to Animas City. He wound up in the jail at Silverton. There a party of masked men overpowered the jail keepers and lynched Wilkinson in his cell. Asked if he had anything to say before his death, Wilkinson said, "Nothing, gentlemen, *adiós!*" [18, 84, 186, 190, 202]

William's Creek Reservoir State Wildlife Area. Built in the 1950s for fishing, this lake has 343 surface acres at 8,241 feet elevation, surrounded by ponderosa pine and spruce forests. Its coldwater fishery of kokanee salmon, rainbow trout, and brook trout is managed by the Colorado Division of Wildlife. William's Creek is three miles west of Pagosa Springs on U.S. 160 and thirty miles north on Forest Service 631. [250]

wolf. The gray wolf that once lived in the basin ranged in North America from the Arctic to Mexico and to both coasts. Although they like a wide variety of habitats, gray wolves are often called timber wolves, perhaps to distinguish them from coyotes or prairie wolves. Wolves once fed on bison, wapiti, deer, and smaller mammals. When cattle and sheep replaced or joined wildlife, wolves naturally accepted the new food source. Their depredations of domestic animals brought them disfavor. With the assistance of various government agencies, they were systematically shot, trapped, and poisoned in the basin, as elsewhere, until they were eradicated, probably in the 1940s. So now, as far as the basin is concerned, the wolf is an extirpated species—an animal that no longer lives in its wild habitat but exists elsewhere. [135]

Wolf Creek Pass. The road over this entrance to the basin was officially opened as the first Colorado auto road over the Continental Divide in August 1916, six years after the highway engineers started building it. Unlike most highway passes, it had no precedent wagon road. Although it experiences fewer snow slides than many other high mountain highways, the road receives the heaviest snowfall in the basin—sometimes as much as seventy feet. In a song made popular by C. W. McCall, a truck driver with a load of chickens lost his brakes on this pass and ended up in downtown Pagosa Springs.

Wolf Creek Pass (elev. 10,850 feet) is twenty-three miles east of Pagosa Springs on U.S. 160. [91, 219]

woodrat. Along with carbon-fourteen and tree-ring dating, archaeologists use woodrat den sites to find out what happened in centuries past. They do this by examining what ancestral "pack rats" stored and what kinds of pollen and other environmental debris stuck to their accumulated urine. The rough, broken terrain of the basin provides ideal habitat for several kinds of these rodents. Desert woodrats grow only twelve inches long while the bushy-tailed mountain variety may get much larger. [135]

Yellow Jacket, Montezuma County, Colorado. A few families settled at springs near here in 1912 and 1913 to try their hand at dryland farming. They managed to raise beans, wheat, and a few other crops while they perhaps cast envious glances at greener fields moistened by the Montezuma Valley Irrigation Company.

Yellow Jacket is eighteen miles north of Cortez on U.S. 666. [50]

Yucca House National Monument. Ancestral Puebloans built shelters and lived at this site during the period 1000–1300. Surrounded by private property, the monument is not publicly accessible but is preserved for future archaeological study. Yucca House is twelve miles south of Cortez. [203]

Appendix A

Entries by Category

Natural Features and Related Sites

Angel Peak National Recreation Site
Animas River
badlands
Baker's Park
Barker Arroyo
Battle Rock
Bisti Wilderness
breaks
canyon
Carson National Forest
chaparral
Chuska Mountains
Cinnamon Pass
Coal Bank Pass
Colorado Plateau
Colorado River
Continental Divide
Cunningham Pass
De-Na-Zin Wilderness
Dolores River
Engineer Pass
Florida River
Hesperus Pass
hogback
Kennebec Pass
La Plata Mountains
La Plata River
Largo Canyon
Lizard Head Pass
Los Piños River
Mancos River
McElmo Canyon
mesa
Molas Pass
Needle Mountains
Ophir Pass

Pagosa Springs
Perins Peak
Piedra River
Pine River
Red Mountain Pass
Rio Grande
rivers
Rocky Mountains
San Juan Basin
San Juan Mountains
San Juan National Forest
San Juan River
Shining Mountains
Shiprock Pinnacle
Sleeping Ute
Stony Pass
Twin Buttes
Ute Pass
vegetation zones
Washington Pass
Wolf Creek Pass

The Native Americans and Ancient Ruins

Anasazi
Anasazi Heritage Center
Ancestral Puebloans
Athapascan
Aztec Ruins National Monument
Capote
Chaco
Chaco Canyon
Chaco Culture National Historic Park
Chimney Rock Archaeological Area
Code Talkers
Domínguez and Escalante Ruins
Falls Creek Archaeological Area

firewater
Geronimo
hogan
Hovenweep National Monument
Indian
Indian reservation
Indian tribe
Jicarilla Apache
Jicarilla Apache Reservation
Lowry Pueblo Ruins
Manuelito
Mesa Verde National Park
Mouache
Narbona
Navajo Nation
Navajo Reservation
Navajo rug
Navajo Tribe
Ouray
peyote
Pueblitos of Dinetah
redskin
Salmon Ruin
Sapiah (Buckskin Charlie)
Southern Ute Reservation
Southern Ute Tribe
trance rite
Uncompahgre
Ute
Ute Mountain Ute Reservation
Ute Mountain Ute Tribe
Weminuche
Yucca House National Monument

The Spanish Influence

Armijo, Antonio
Coronado, Francisco Vásquez de
Domínguez and Escalante Expedition
El Camino Real
horse
Mexico
Old Spanish Trail
Oñate, Juan de
Pueblo Revolt
Rio Abajo
Rio Arriba

Rivera, Juan María de
Spanish and Mexican land grants
Tierra Amarilla Land Grant

The Anglo-American Frontier

Baker, Charles
barbed wire
blacklist rule
Bosque Redondo
Brunot Agreement
Buffalo Soldiers
burro
Carson, Kit
cattle
cattle brands
cattle drives
cattlemen-sheepherders wars
cattle trails
Civil War
Code of the West
Coe, George
Concord Coach
Conestoga wagon
cowboys
dance hall
Deseret
dogie
dragoons
dry-gulch
dude
dugout
fanning
fast draw
fence wars
fort
Fort Defiance
Fort Lewis
Fort Lowell
Fremont, John Charles
general store
Goodnight, Charles
Graves, Alf
gunfight
gunfighting relatives
Hamlet Cabin
Hayden Survey

Hendrickson, William P.
homesteading
Jackson, William Henry
law officer
Lincoln County War
longhorn
lynching
Manifest Destiny
marshal
maverick
McCarty, Henry "Billy the Kid"
Mears, Otto
Meeker Massacre
Moorman, Henry Reed
Mormons
mule
mustang
Nance, Tom
nester
Palmer, William Jackson
pinto
prostitution
red light district
Remington, Frederic
roundup
rustling
saloon
Santa Fe Trail
Sheek, James Lorenz
sheep
Sheridan, Philip Henry
Sherman, William Tecumseh
Sherman Silver Purchase Act
Stockton, Ike
Stockton, Port
Sullivan, James J.
trading post
trapper
Twin Crossings
Ute Strip
vigilantes
Watermelon Bust
Wilkinson, Bert

**Communities, Towns, and Counties:
Past and Present**

Alcatraz
Allison
Amargo
Animas City
Animas Forks
Arboles
Archuleta County
Arriola
Aztec
Bayfield
Beklabito
Big Bend
Blanco
Blanco Trading Post
Bloomfield
Bondad
Breen
Burnham
Cahone
Carbon Junction
Cascade
Cedar Hill
Chromo
Cortez
Counselor
Crystal
Dolores
Dolores County
Dove Creek
Dulce
Dunton
Durango
Edith
Eureka
Falfa
Farmington
Flora Vista
Florida
Fruitland
Gavilan

Gem Village
Gladstone
Gobernador
Hermosa
Hesperus
Hinsdale County
Howardsville
Ignacio
Junction City
Kirtland
Kline
La Baca
La Plata
La Plata City
La Plata County
La Posta
Lebanon
Lewis
Lumberton
Mancos
Marvel
Mayday
Mineral County
Monero
Montezuma County
Nageezi
Naschitti
Newcomb
Oxford
Pagosa Junction
Pagosa Springs
Parrott City
Pleasant View
Redmesa
Rico
Rio Arriba County
Rockwood
San Juan Counties
San Juan County (Colorado)
San Juan County (New Mexico)
Sandoval County
Sanostee
Sheep Springs
Shiprock
Silverton
Stoner
Taos County
Tiffany

Toadlena
Tocito
Towaoc
Turley
Two Gray Hills
Waterflow
Yellow Jacket

Places, Features, and Organizations

Agricultural Science Center at
    Farmington
Alpine Loop
Alpine Triangle Recreation Area
Anasazi Recreation Area
Andrew's Lake State Wildlife Area
Animas–La Plata Project
Baker's Bridge
Bodo State Wildlife Area
Bureau of Indian Affairs
Bureau of Land Management
Bureau of Reclamation
Colorado (State)
Colorado–New Mexico border
Colorado River Compact
Colorado River Storage Project
Colorado Territory
Colorado Trail
counties
Crow Canyon Archaeological Center
Denver and Rio Grande Railway
Devil Creek State Wildlife Area
Dolores Lumber Railroads
Dolores Project
Dolores River Recreation Area
Domínguez and Escalante Memorial
    Highway
Durango and Silverton Narrow
    Gauge Railroad
Durango Fish Hatchery
Echo Canyon Reservoir State Wildlife
    Area
Electra Lake
Fish Creek State Wildlife Area
Florida Project
Fort Lewis College
Four Corners

Four Corners Generating Station
Four Corners Monument Navajo
   Tribal Park
Groundhog Reservoir State Wildlife
   Area
Hammond Project
Haviland Lake State Wildlife Area
historic places
Jackson Lake State Wildlife Area
Joe Moore Reservoir State Wildlife
   Area
La Plata Mine
Lizard Head Wilderness
Lone Dome State Wildlife Area
Mancos Project
Mancos State Recreation Area
McPhee Reservoir
Milagro Co-Generation Facility
Million Dollar Highway
Morgan Lake
motto
Narraguinnep Reservoir State Wildlife
   Area
national forest system
national park system
natural gas
Navajo Indian Irrigation Project
Navajo Lake State Park
Navajo Lake State Recreation Area
Navajo Mine
Navajo Reservoir
Navajo Reservoir Management Area
Navajo State Wildlife Area
Navajo Trail
New Mexico
New Mexico Territory
oil
108th Meridian
Pastorius Reservoir State Wildlife
   Area
Perins Peak State Wildlife Area
Pine River Project
Puett Reservoir State Wildlife Area
railroads
Red Apple Flyer
Rio Grande and Pagosa Springs
   Railroad

Rio Grande and Southwestern
   Railroad
Rio Grande, Pagosa and Northern
   Railroad
Rio Grande Southern Railroad
San Juan Basin Area Vocational-
   Technical School
San Juan Basin Research Center
San Juan College
San Juan Generating Station
San Juan Mine
San Juan River Highway Bridge
San Juan Skyway
scenic and historic byways
Silverton, Gladstone and Northerly
   Railroad
Silverton Northern Railroad
Silverton Railroad
Simon Canyon Recreation Area
South San Juan Wilderness
state parks and recreation areas
state trust lands
Summit Reservoir State Wildlife Area
Tacoma
territory
Totten Reservoir State Wildlife Area
Trail of the Ancients
Trimble Springs
Upper Colorado River Basin Compact
uranium
Ute Mountain Tribal Park
Weminuche Wilderness
wildlife areas
William's Creek Reservoir State Wild-
   life Area

Plants and Animals

bear
beaver
bighorn
blue spruce
buffalo (bison)
chipmunk
cottonwood
coyote
deer

Douglas fir
eagle
Engelmann spruce
Gambel oak
larkspur
leafy spurge
marmot
mountain lion
piñon
ponderosa pine
prairie dog
quaking aspen
roadrunner
Rocky Mountain juniper
squirrel
subalpine fir
tree species
tumbleweed
wapiti (elk)
weed species
wildflower species
wildlife
wolf
woodrat

Other People and Subjects

Armstrong, Neil Alden
Cather, Willa
Dempsey, Jack
Elitch, Mary
Grey, Zane
Harman, Fred
Hillerman, Tony
Kroeger, Frederick Wilson
L'Amour, Louis
motto
movies
museums
Out West
populations
Rocky Mountain oyster
rodeo
Rogers, Will
Tabor, Horace A. W.
Tison Gang
unidentified flying object (UFO)

Appendix B

Suggested Itineraries

A tour in the San Juan Basin reveals many points of interest, a few marked by signs, some otherwise apparent. But the basin conceals many stories, and some features hide from the traveler. This list groups entries about happenings, communities, and features—both self-evident and hidden—on or near the basin's main routes. To use this guide, pick a starting point. Then read down the list toward the destination. If your starting point is listed as a destination, start at the bottom of the list and read up.

(A) Bloomfield to Counselor via New Mexico 44

Bloomfield
Angel Peak National Recreation Area
De-Na-Zin Wilderness
Blanco Trading Post
Nageezi
Chaco Cultural National Historic Park
Chaco Canyon
Counselor

(B) Chimney Rock area to U.S. 160 and Colorado 172 intersection via Ignacio

Chimney Rock Pinnacle
Chimney Rock Archaeological Area
Navajo Lake State Recreation Area
Navajo Reservoir Management Area
Navajo State Wildlife Area
Arboles
Allison
Tiffany
LaBaca
Ignacio
Southern Ute Tribe
Capote
Mouache
Southern Ute Reservation

Oxford
Pastorius Reservoir State Wildlife
    Area
Falfa

(C) Cortez to Dove Creek via U.S. 666

Cortez
Anasazi Recreation Area
Hovenweep National Monument
Crow Canyon Archaeological Center
Arriola
Lebanon
Narraguinnep Reservoir State Wildlife
    Area
Anasazi Heritage Center
Domínguez and Escalante Ruins
McPhee Reservoir
Dolores Project
Lewis
Yellow Jacket
Pleasant View
Lowry Pueblo Ruins
Cahone
Lone Dome State Wildlife Area
Dolores River Recreation Area
Grey, Zane
Dove Creek

San Juan Basin Itineraries

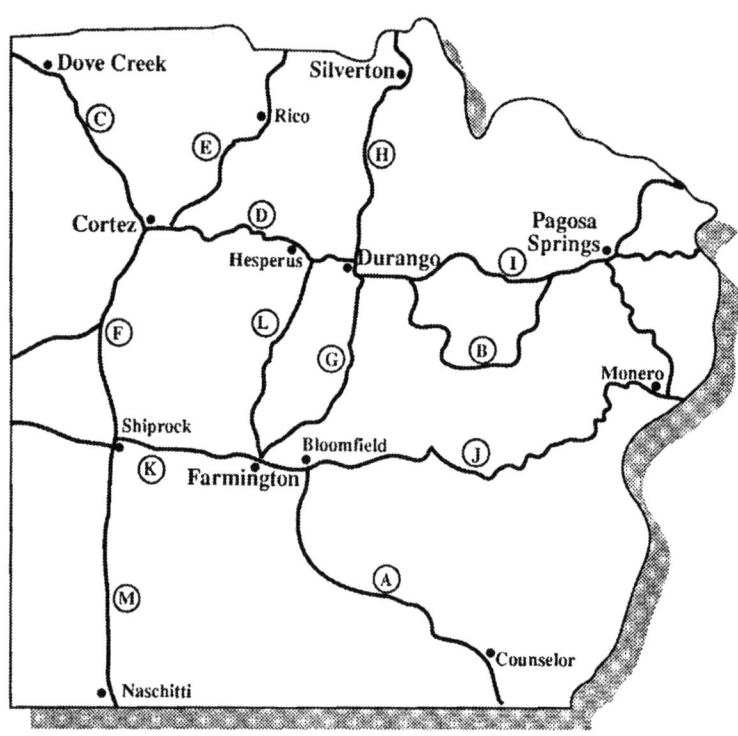

A. Bloomfield to Counselor via New Mexico 44
B. Chimney Rock Area to U.S. 160 and Colorado 172 intersection via Ignacio
C. Cortez to Dove Creek via U.W. 666
D. Cortez to Durango via U.S. 160
E. Cortez to Rico via Colorado 145
F. Cortez to Shiprock via U.S. 666
G. Durango to Farmington via U.S. 550
H. Durango to Red Mountain Pass via U.S.550
I. Durango to Wolf Creek Pass via U.S. 160
J. Farmington to Monero via U.S. 64
K. Farmington to Shiprock via U.S. 64
L. Farmington to Hesperus via New Mexico 170, Colorado 140
M. Naschitti to Shiprock via U.S. 666

(D) Cortez to Durango via U.S. 160

Cortez
Totten Reservoir State Wildlife Area
San Juan Basin Area Vocational-
    Technical School
Mesa Verde National Park
Mancos
Cather, Willa Sibert
Summit Reservoir State Wildlife Area
Sheek, James L.
Domínguez and Escalante Memorial
    Highway
Rio Grande Southern Railroad
Mancos Project
Mancos State Recreation Area
Joe Moore Reservoir State Wildlife
    Area
Summit Reservoir State Wildlife Area
Puett Reservoir State Wildlife Area
Hesperus
Parrott City
Mayday
La Plata City
La Plata Mountains
Bodo State Wildlife Area
Perins Peak State Wildlife Area
Perins Peak
Rogers, Will
Tabor, Horace A. W.
Durango

(E) Cortez to Rico via Colorado 145

Cortez
Big Bend
Dolores
Dolores Lumber Railroads
Dolores River
Dolores River Recreation Area
Stoner
Fish Creek State Wildlife Area
Groundhog Reservoir State Wildlife
    Area

Rico
Dunton
Lizard Head Wilderness
Lizard Head Pass

(F) Cortez to Shiprock via U.S. 666

Cortez
McElmo Canyon
Battle Rock
Ute Mountain Ute Tribe
Ute Mountain Ute Reservation
Weminuche
Sleeping Ute Mountain
Yucca House National Monument
Towaoc
Ute Mountain Tribal Park
Four Corners
Four Corners Monument Navajo
    Tribal Park
Shiprock Pinnacle
Shiprock

(G) Durango to Farmington via U.S. 550

Durango
gunfight
L'Amour, Louis
Kroeger, Frederick Wilson
Animas–La Plata Project
Carbon Junction
Red Apple Flyer
La Posta
Bondad
Twin Crossings
Cedar Hill
Aztec
Hamlet Cabin
Aztec Ruins National Monument
Flora Vista
Farmington
Watermelon Bust
Coe, George Washington

(H) Durango to Red Mountain Pass via U.S. 550

Durango
Durango Fish Hatchery
Durango and Silverton Narrow Gauge
  Railroad
Colorado Trail
Falls Creek Archaeological Area
Animas City
Trimble Springs
Hermosa
Baker's Bridge
Kennebec Pass
Rockwood
Haviland Lake State Wildlife Area
Electra Lake
Armstrong, Neil Alden
Tacoma
Needle Mountains
Cascade
Coal Bank Pass
Andrew's Lake State Wildlife Area
Molas Pass
Weminuche Wilderness
Silverton
Baker, Charles
Alpine Loop
Silverton, Gladstone and Northerly
  Railroad
Gladstone
Baker's Park
Cunningham Pass
Howardsville
Eureka
Stony Pass
Silverton Northern Railroad
Cinnamon Pass
Animas Forks
Engineer Pass
Alpine Triangle Recreation Area
Silverton Railroad
Ophir Pass
Million Dollar Highway
Red Mountain Pass

(I) Durango to Wolf Creek Pass via U.S. 160

Durango
Dempsey, Jack
Elitch, Mary
Fort Lewis College
Florida
Florida River
Florida Project
Gem Village
Bayfield
Pine River Project
Piedra River
Chimney Rock
Tison Gang
Devil Creek State Wildlife Area
Harman, Fred
Pagosa Springs
Pagosa Springs (town)
Rio Grande, Pagosa and Northern
  Railroad
Echo Canyon Reservoir State Wildlife
  Area
Edith
William's Creek Reservoir State Wild-
  life Area
South San Juan Wilderness
Wolf Creek Pass

(J) Farmington to Monero via U.S. 64

Farmington
San Juan College
Junction City
Salmon Ruin
Bloomfield
Milagro Co-Generating Facility
Blanco
Alcatraz
Hammond Project
Largo Canyon
Pueblitos of Dinetah
Turley
Navajo Reservoir

Navajo Lake State Park
Simon Canyon Recreation Area
Gobernador
Carson National Forest
Jicarilla Apache
Jicarilla Apache Reservation
Dulce
Lumberton
Rio Grande and Pagosa Springs
    Railroad
Rio Grande and Southwestern
    Railroad
Amargo
Tierra Amarilla Land Grant
Chromo
Monero

(K) Farmington to Shiprock via U.S. 64

Farmington
unidentified flying object (UFO)
Twin Buttes
Agricultural Science Center at
    Farmington
Navajo Indian Irrigation Project
Kirtland
Fruitland
Four Corners Generating Station
Morgan Lake
Navajo Mine
San Juan Generating Station
San Juan Mine
Waterflow
Hillerman, Tony
hogback
Shiprock

(L) Farmington to Hesperus via New Mexico 170 and Colorado 140

Farmington
Jackson Lake State Wildlife Area
Barker Arroyo
La Plata
La Plata Mine
Redmesa
Marvel
Kline
Breen
San Juan Basin Research Center
Fort Lewis
Hesperus Pass
Hesperus

(M) Naschitti to Shiprock via U.S. 666

Naschitti
Chuska Mountains
Washington Pass
trading post
Sheep Springs
Crystal
Newcomb
Two Gray Hills
Toadlena
Burnham
Tocito
Sanostee
Beklabito
San Juan River Highway Bridge
Shiprock

Appendix C

Names of Places and Features

By 1848, when the United States took over the San Juan Basin and the rest of the vast territory given up by Mexico at the end of the Mexican-American War, the Native Americans, Spaniards, and Mexicans had already named many of its natural features. The Anglo Americans adopted many of those names and imparted some of them to their towns and counties. Other titles came from a variety of sources; in some cases, we know what a name means but we don't know who first applied it. To avoid redundancy, this list omits duplicate feature names, place-names that repeat feature names, and names of unknown origin.

Allison. For Allison Stocker, who surveyed and helped develop southeastern La Plata County.

Animas River. The Spanish called it El Rio de las Animas Perdidas, which means "the river of the souls lost in purgatory," or "the river of lost souls." (They gave the same name to a river in southeastern Colorado.)

Archuleta County. When the legislature created the county out of Canejos County, they named it to honor state senator Antonio D. Archuleta.

Arboles. From *arbolado,* Spanish for "tree-lined"—a reference to the woods along the Piedra River.

Arriola. An early Spaniard general.

Aztec Ruins. Named from the false notion that they were remnants of Central America's ancient Aztec civilization.

Baker's Park. For Charles Baker, who led 1860s prospecting parties into the San Juan Mountains.

Barker Arroyo. For Aaron Barker, who was murdered in this gulch.

Bayfield. For W. A. and Laura E. Bay, who owned the land that is now the townsite.

Blanco. Spanish for "white," like the silica in the nearby hills.

Bloomfield. An early Mormon settler who got church money to finance an irrigation system.

Bondad. Spanish for "goodness" or "helpfulness."

Breen. For Dr. Thomas H. Breen, who homesteaded here.

Burnham. For trader Roy Burnham.

Cahone. From the Spanish *cajón,* meaning "big box," named by an early postmaster for a nearby box canyon.

Carson National Forest. For Kit Carson.

Cedar Hill. A common name for a knoll with cedarlike junipers.

Chimney Rock. The Spanish called this conspicuous feature *piedra,* meaning "rock," or *piedra parado,* "standing rock," but the Anglo Americans called it what it looks like.

Chromo. A Greek word for "color," like the surrounding landscape.

Chuska Mountains. A corruption of *shashgai*, Navajo for "white spruce."

Colorado River. In Spanish Rio Colorado means "red river," a term Juan de Oñate first used to describe the stream we call the Little Colorado River in Arizona.

Cortez. Named for the Spanish conqueror, but not by the Spaniards; the name was suggested by an Anglo homesteader.

Counselor. For Jim Counselor, a trader.

Crystal. A corruption of *Tonlt'ili*, Navajo for "where crystal water flows out."

Cunningham Creek. For Major W. H. Cunningham, who brought Chicago mine investors with him to the San Juans.

De-Na-Zin. Navajo for "standing creek."

Dolores River. The Spanish called it Rio de Nuestra Senora de las Dolores, or "River of Our Lady of Sorrows."

Dove Creek. A pioneer freighter saw many doves at a nearby creek.

Dulce. Spanish for "sweet"—a reference to the water.

Durango. Colorado territorial governor A. C. Hunt named the town after visiting Durango, Mexico. The term may come from the Basque words *ur*, meaning "water," and *ango*, meaning "villa" (the "D" was added by the Spanish).

Edith. Named by lumberman Edgar M. Biggs for his baby daughter.

Electra Lake. This reservoir stores water used to generate electricity.

Farmington. The name reflects the community's farming heritage.

Flora Vista. Spanish for "flower view."

Florida River. Spanish for "flowering." Escalante named it.

Fort Lewis. For Lieutenant Colonel William H. Lewis, who was killed fighting the Cheyennes in Kansas. (He was a descendant of explorer Meriwether Lewis.)

Fruitland. A fruit-tree-nurturing postmaster named this community.

Gavilan. Spanish for "sparrow hawk."

Gem Village. For a 1940s colony of gem and mineral artisans.

Gobernador. Spanish for "governor."

Hermosa. Spanish for "beautiful." The name was applied by settlers in the 1870s.

Hesperus Mountain. The Hayden survey party took this name from Henry Wadsworth Longfellow's poem, "The Wreck of the Hesperus." It comes from the Greek word for "evening star."

Hinsdale County. Named in honor of George A. Hinsdale, who served as lieutenant governor of Colorado.

Hovenweep. A Ute term for "deserted valley."

Howardsville. After prospector George W. Howard.

Ignacio. For Ignacio, a Ute chief, and St. Ignatius, a Christian martyr.

Jackson Butte. For William Henry Jackson, famous photographer of the West.

Kirtland. Named by its Mormon founders for one of their Ohio settlements. (An air force base in Albuquerque with the same title was named for a pilot.)

Kline. The name came from the Mormons who settled there.

La Baca. Likely a corruption of Spanish *la boca*, "the mouth," referring to the mouth of Los Piños River.

La Plata: Spaniards found the mountains to yield *plata*, or "silver," and named them Sierra de la Plata. *Sierra* is Spanish for "saw." In this case it refers to the rugged, sawlike peaks.

La Posta. In Spanish, *posta* means "stage station," which this place was.

Largo Canyon. *Largo* is Spanish for "long."

Lebanon. Pioneers thought the town was built in a forest of cedars, like the biblical Lebanon, although the trees are junipers.

Lightner Creek. For E. C. Lightner.

Lizard Head Pass. A nearby peak looks like its name.

Lowry Ruins. After homesteader George Lowry.

Lumberton. Started with the lumber industry.

Mancos River. From the Spanish word *manco*, meaning "crippled."

Marvel. After the Marvel Midget flour mill that ground grain for the pioneers.

Mayday. The spring festival or the distress signal?

McElmo Canyon. A prospector by this name died there.

Mesa Verde. Spanish for "green table."

Monero. Spanish for "dark" or "swarthy."

Montezuma County. Some settlers mistakenly thought the ruins in the area were remnants of the Central American Aztec civilization, so they named the county after the famous Aztec ruler.

Nageezi. Means "squash" in Navajo.

Naschitti. In the Navajo language, this means "badger water" or "scratching for water."

Newcomb. Author J. Newcomb and his wife, Franc, ran a trading post at this location.

New Mexico. In the language of the Central American Aztecs, *mexico* meant "place of the moon." This may be where the name came from, but it could also have come from the Aztec war god, Mexitl. In 1562 a Mexican governor first referred to the northern part of New Spain as "another" or "new" Mexico.

Ophir. The biblical name for the location of King Solomon's mines.

Parrott City. For the San Francisco banking family that financed the original mining camp.

Pagosa Springs. *Pagosa* is a Ute word meaning "healing waters."

Perins Peak. After Charles M. Perins, who surveyed the Durango townsite.

Piedra River. Derived from Rio de la Piedra Parada, Spanish for "River of the Standing Rock," which may refer to nearby Chimney Rock or to the river's canyon walls.

Pine River. From the Spanish Rio de los Piños, meaning "River of the Pines."

Red Mountain. For the iron oxide on its sides.

Rico. This word means "rich" in Spanish. The place was named by the prospectors who found silver nearby.

Rio Arriba. Spanish for "river higher up," referring to the upper Rio Grande.

Rio Grande. Spanish for "large river."

Rockwood. Named for Thomas Rockwood, son of Sir Richard Rockwood of Boston, the "sir" reflecting his British royalty.

Rocky Mountains. The name is a translation from the French Montagnes Rocheuses, or "Rocky Mountains," used by the French Canadians in reference to the Assiniboin native tribe.

Salmon Ruin. It is on land home-steaded by George Salmon.

Sandoval County. For the Spanish settlers and cattlemen of that name whose lineage goes back to Juan de Díos Sandoval Martínez of Mexico City.

San Juan. A name applied by Spaniards for Saint John the Baptist.

Sanostee. This is a Navajo word. It may mean "rocks around."

Shiprock. The Navajos call this prominent feature Tse Bit'a'i, meaning "rock with wings" or "winged rock." For Anglo Americans, this peak is named for what it looks like.

Silverton. From "silver town." Also rumored to have come from a miner who said, "We have no gold, but a ton of silver."

Stoner. After Stony Creek, so named for its rocky bed.

Tacoma. The generators shipped there were made, and labeled, for a plant at Tacoma, Washington.

Tierra Amarilla. Spanish for "yellow earth."

Toadlena. A corruption of *tohaali*, Navajo for "water bubbling up."

Tocito. Navajo for "warm springs."

Towaoc. When a Ute Mountain subagency was moved there, the Utes said it was *towaoc*, meaning "all right."

Trimble Springs. For W. C. Trimble, who settled there.

Turley. An early postmaster.

Twin Crossings. Where the Animas River forms a horseshoe, the pioneers had to cross it twice.

Vallecito. Spanish for "little valley."

Washington Pass. For Colonel James Macrae Washington, who led an early expedition into the Navajo's homeland.

Windom Peak. For William Windom, secretary of the treasury under President James Garfield.

Yellow Jacket. Wasp nests in a canyon prompted this name.

Sources

Books and articles about western Americana and this work's other subjects seem limitless. And since the San Juan Basin is an exciting part of the West (that's one reason for this book), places within the basin don't suffer from a lack of source material either. For the West as a whole, this list includes only a fraction of the material available. For the basin, however, it includes most of the reliable (and some not so reliable) works that are useful and available.

To learn more about the West, you might start with *The Oxford History of the American West*, edited by Clyde A. Milner II, Carol A. O'Connor, and Martha A. Sandweiss. To focus on the Southwest, try *The Southwest* by David Lavender.

Since a definitive history of the San Juan Basin is yet to be written, you'll have to consult a larger selection to take an in-depth look at that smaller region. After being out of print for several decades, *Pioneers of the San Juan Country* was reprinted in 1995. Its collection of articles provide good insights, as does *The San Juan Basin: My Kingdom Was a County*, by Eleanor MacDonald and pioneer John B. Arrington. Unfortunately, Arrington's biography was privately published and is not readily availabile. You may find it is on basin library shelves. Beyond these selections, many listed publications cover specific basin places and events. The source's number appears at the end of each article for which the source was used, so you can pursue a subject further.

Manuscripts

1. Cornelius, Oliver F. "Pioneer History and Reminisces of the San Juan Basin," 1932. Assorted Writings on the History of Durango, Colorado, Durango Public Library.
2. Culhane, Albert E. "A History of the Settlement of La Plata County, Colorado." M.A. thesis, University of Colorado, 1934.
3. Duke, Robert W. "Political History of San Juan County, New Mexico." M.A. thesis, University of New Mexico, 1947.
4. Dwyer, Robert. Interview, March 12, 1887. Bancroft Library, University of California, Berkeley.
5. "Forest History, Vol. 1, 1905–1971, San Juan and Montezuma Forests, Colorado," n.d. Durango Public Library.
6. McGinn, Elinor M. "Sixty Years on the Durango-Farmington Branch Railroad (1905–1965)." n.d. Fort Lewis College Library, Southwest Section.
7. Sumner, George T. Untitled manuscript, 1932. Assorted Writings on the History of Durango, Colorado, Durango Public Library.

Sources

Newspapers

8. *Denver Post.*
9. *Denver Republican.*
10. *Dolores Rico News.* Quoted in F. [Francis Louis Crocchiola] Stanley, *The Private War of Ike Stockton* (Denver: World Press, 1959).
11. *Durango Herald.*
12. *Durango Herald Democrat.*
13. *Durango Record.*
14. *Durango Weekly Herald.*
15. *Farmington Daily Times.*
16. *Farmington Times Hustler.*
17. *Rocky Mountain News.*
18. *San Juan Herald.*

Books

19. Abbott, Carl, Stephen J. Leonard, and David McComb. *Colorado: A History of the Centennial State.* Rev. ed. Boulder: Colorado Associated University Press, 1982.
20. Adams, Ramon F. *The Cowman and His Code of Ethics.* Austin: Encino Press. 1969.
21. Albright, Horace M., Russell E. Dickenson, and William Penn Mott Jr. *National Park Service: The Story behind the Scenery.* Las Vegas, Nev.: KC Publications, 1987.
22. Athearn, Robert G. *Rebel of the Rockies: The Denver and Rio Grande Western Railroad.* New Haven: Yale University Press, 1962.
23. ———. *William Tecumseh Sherman and the Settlement of the West.* Norman: University of Oklahoma Press, 1956.
24. Ayer, Eleanor H. *The Anasazi.* New York: Walker and Company, 1993.
25. Ball, Larry D. *Desert Lawmen: The High Sheriffs of New Mexico and Arizona, 1846–1912.* Albuquerque: University of New Mexico Press, 1992.
26. ———. *The United States Marshals of New Mexico and Arizona Territories, 1846–1912.* Albuquerque: University of New Mexico Press, 1978.
27. Bear, Leith Lende. *Trimble Hot Springs.* 2d ed. Durango: Trimble Hot Springs, 1985.
28. Beck, Warren A. *New Mexico: A History of Four Centuries.* Norman: University of Oklahoma Press, 1962.
29. Bird, Allan G. *Bordellos of Blair Street: The Story of Silverton, Colorado's Red Light District.* Grand Rapids: The Other Shop, 1987.
30. Briggs, Walter. *Without Noise of Arms: The 1776 Dominguez-Escalante Search for a Route from Santa Fe to Monterey.* Flagstaff, Ariz.: Northland Press, 1976.
31. Carlson, Alvar W. *The Spanish-American Homeland: Four Centuries in New Mexico's Rio Arriba.* Baltimore: Johns Hopkins University Press, 1990.

32. Carstensen, Vernon R., ed. *The Public Lands: Studies in the History of the Public Domain.* Madison: University of Wisconsin Press, 1963.
33. Cerquone, Joseph. *In Behalf of the Light: The Dominguez and Escalante Expedition, 1776.* Denver: Dominguez Escalante Bicentennial Expedition, 1976.
34. Clarke, James W. *Last Rampage: The Escape of Gary Tison.* Boston: Houghton Mifflin, 1988.
35. Cleland, Robert G. *This Reckless Breed of Men: The Trappers and Fur Traders of the Southwest.* New York: Alfred A. Knopf, 1963.
36. Coan, Charles F. *The County Boundaries of New Mexico.* Santa Fe: New Mexico Legislative Council, 1965.
37. Coe, George W. *Frontier Fighter: The Autobiography of George W. Coe Who Fought and Rode with Billy the Kid.* Boston: Houghton Mifflin, 1934.
38. Colorado Weed Management Association. *Colorado's 30 Troublesome Weeds: Plants That Threaten Our Natural Resources.* 2d ed. Fort Collins: Colorado Weed Management Association, 1993.
39. Crum, Josie M. *Three Little Lines: Silverton Railroad; Silverton Gladstone & Northerly; Silverton Northern.* Durango: Durango Herald News, 1960.
40. Dallas, Sandra. *Colorado Ghost Towns and Mining Camps.* Norman: University of Oklahoma Press, 1990.
41. DeHaan, Vici. *State Parks of the West: America's Best-Kept Secrets.* Evergreen, Colo.: Cordillera Press, 1990.
42. Delaney, Robert W. *The Ute Mountain Utes.* Albuquerque: University of New Mexico Press, 1989.
43. Dobie, James F. *The Longhorns.* New York: Barnhall House, 1941.
44. Ebright, Malcolm. *The Tierra Amarilla Grant: A History of Chicanery.* Santa Fe: Center for Land Grant Studies, 1980.
45. Egan, Ferol. *Fremont: Explorer for a Restless Nation.* Garden City, N.Y.: Doubleday, 1977.
46. Fisher, John S. *A Builder of the West: The Life of General William Jackson Palmer.* Caldwell, Idaho: Caxton Printers, 1939.
47. Flanagan, Mike. *Out West.* New York: Harry N. Abrams, 1987.
48. Fort Lewis Mesa Reunion History Committee. *History of Southwestern La Plata County in Colorado.* Durango: Fort Lewis Mesa Reunion History Committee, 1991.
49. ———. *Pioneers of Southwest La Plata County, Colorado.* Bountiful, Utah: Family History Publishers, 1994.
50. Freeman, Ira S. *A History of Montezuma County, Colorado.* Boulder: Johnson Publishing, 1958.
51. Furman, Agnes M. *Tohta: An Early Day History of the Settlement of Farmington and San Juan County, New Mexico 1875–1900.* Wichita Falls, Tex.: Nortex Press, 1977.
52. Gilbert, Bil. *The Trailblazers.* 3d ed. The Old West Series. Alexandria, Va.: Time-Life Books, 1979.

53. Goff, Richard, Robert H. McCaffree, and Doris Sterbenz. *Centennial Brand Book of the Colorado Cattlemen's Association*. Denver: Colorado Cattlemen's Centennial Commission, 1967.

54. Goff, Richard, and Robert H. McCaffree. *Century in the Saddle: the 100 Year History of the Colorado Cattlemen's Association*. Denver: Colorado Cattlemen's Centennial Commission, 1967.

55. Gomez, Arthur R. *Quest for the Golden Circle: The Four Corners and the Metropolitan West, 1945–1970*. Albuquerque: University of New Mexico Press, 1994.

56. Hafen, LeRoy R., and Ann W. Hafen. *Fremont's Fourth Expedition: A Documentary Account of the Disaster of 1848–49*. Glendale: Arthur H. Clark Co., 1960.

57. ———. *Old Spanish Trail: Santa Fe to Los Angeles*. Glendale: Authur H. Clark Co., 1954.

58. Hafen, LeRoy R., ed. *Mountain Men and Fur Traders of the Far West*. Lincoln: University of Nebraska Press, 1982.

59. Haley, J. Evetts. *Charles Goodnight: Cowman and Plainsman*. New ed. Norman: University of Oklahoma Press, 1949.

60. Hales, Peter B. *William Henry Jackson and the Transformation of the American Landscape*. Philadelphia: Temple University Press, 1988.

61. Haley, James L. *Apaches: A History and Cultural Portrait*. Garden City, N.Y.: Doubleday, 1981.

62. Hart, John L. J. *Fourteen Thousand Feet: A History of the Naming and Early Ascents of the High Colorado Peaks*. 2d ed. Denver: Colorado Mountain Club, 1931.

63. Hilton, George W. *American Narrow Gauge Railroads*. Stanford, Calif.: Stanford University Press, 1990.

64. Hoffman, Virginia. *Navajo Biographies*. Vol. 1. Phoenix: Navajo Curriculum Center Press, 1974.

65. Howe, Elvon L., ed. *Rocky Mountain Empire*. Garden City, N.Y.: Doubleday, 1950.

66. Hunt, Corrine, and Jack Gurtler. *The Elitch Garden Story*. Denver: Rocky Mountain Writers Guild, 1982.

67. Hutton, Paul A. *Phil Sheridan and His Army*. Lincoln: University of Nebraska Press, 1985.

68. Jackson, William H. *Time Exposure: An Autobiography of William Henry Jackson*. 1940. Reprint, with a foreword by Ferenc M. Szasz, Albuquerque: University of New Mexico Press, 1986.

69. Jackson, William H., and William H. Holmes. *Mesa Verde and the Four Corners: Hayden Survey, 1874–1876*. Ouray, Colo.: Bear Creek Publishing, 1981.

70. Jacobs, Randy. *A Colorado High: The Official Guide to the Colorado Trail*. Englewood, Colo.: Colorado Trail Foundation/Westcliff, 1992.

71. James, H. L. *Posts and Rugs: The Story of Navajo Rugs and Their Homes*. Globe, Ariz.: Southwest Parks and Monuments Association, 1976.

72. James, L. F., et al. *Plants Poisonous to Livestock in the Western States.* Washington, D.C.: U.S. Department of Agriculture, 1982.

73. Jarvis, Marion. *"Come on in Dearie" or Prostitutes of Early Durango.* Durango: Durango Herald, 1976.

74. Jefferson, James, Robert W. Delaney, and Gregory C. Thompson. *The Southern Utes: A Tribal History.* 2d ed. Edited by Floyd A. O'Neil. Ignacio, Colo.: Southern Ute Tribe, 1973.

75. Jessen, Kenneth C. *Colorado Gunsmoke: True Stories of Outlaws and Lawmen on the Colorado Frontier.* Boulder: Pruett Publishing, 1986.

76. John, Elizabeth A. H. *Storms Brewed in Other Men's Worlds: The Confrontation of Indians, Spanish, and French in the Southwest, 1540–1795.* College Station: Texas A&M University Press, 1975.

77. Josephy, Alvin M., Jr., ed. *War on the Frontier (The Civil War).* Alexandria Va.: Time-Life Books, 1986.

78. Lamar, Howard R. *The Far Southwest, 1846–1912: Territorial History.* New Haven: Yale University Press, 1966.

79. Lanner, Ronald M. *The Pinon Pine: A Natural and Cultural History.* Reno, Nev.: University of Nevada Press, 1981.

80. Lavender, David. *The Southwest.* New York: Harper & Row, 1980.

81. Little, Elbert L. *The Audubon Society Field Guide to North American Trees: Western Region.* New York: Alfred A. Knopf, 1980.

82. Locke, Raymond F. *The Book of the Navajo.* 5th ed. Los Angeles: Mankind Publishing, 1992.

83. Lomax, John A., and Alan Lomax. *Cowboy Songs and Other Frontier Ballads.* New York: Macmillan, 1986.

84. Luzar, Retha B. *The Animas City Story: A Forerunner of Durango, Colorado.* N.p., 1978.

85. MacDonald, Eleanor D., and John B. Arrington. *The San Juan Basin: My Kingdom Was a County.* Denver: Green Mountain Press, 1970.

86. Marsh, Charles S. *People of the Shining Mountains: The Utes of Colorado.* Boulder: Pruett Publishing, 1982.

87. McNitt, Frank. *The Indian Traders.* Norman: University of Oklahoma Press, 1962.

88. Miller, Floyd. *Bill Tilghman: Marshal of the Last Frontier.* Garden City, N.Y.: Doubleday, 1968.

89. Milner, Clyde A., II, Carol A. O'Connor, and Martha A. Sandweiss, eds. *The Oxford History of the American West.* New York: Oxford University Press, 1994.

90. Morgan, Dale L. *The State of Deseret.* Logan, Utah: University of Utah Press/Utah Historical Society, 1987.

91. Motter, John M. *Pagosa Country: The First Fifty Years.* Pagosa Springs, Colo.: J. M. Motter, n.d.

92. Myrick, David. F. *New Mexico Railroads: An Historical Survey.* Rev. ed. Albuquerque: University of New Mexico Press, 1990.

93. Newcomb, Franc J. *Navajo Neighbors.* Norman: University of Oklahoma Press, 1966.

94. Nolan, Frederick W. *The Lincoln County War: A Documentary History.* Norman: University of Oklahoma Press, 1992.
95. Noland, James M., and Robert L. Beers. *History of the Electra Lake Sporting Club.* Durango: Electra Lake Sporting Club, 1994.
96. Nossaman, Allen. *Many More Mountains.* Vol. 1. Denver: Sundance Publications, 1989.
97. ———. *Many More Mountains.* Vol. 2. Denver: Sundance Publications, 1993.
98. O'Neal, Bill. *Cattlemen vs. Sheepherders: Five Decades of Violence in the West, 1880–1920.* Austin: Eakin Press, 1989.
99. Onis, Jose de, ed. *The Hispanic Contribution to the State of Colorado.* Boulder: Westview Press, 1976.
100. Ormes, Robert M. *Tracking Ghost Railroads in Colorado.* Colorado Springs: Century One Press, 1980.
101. Osterwald, Doris B. *Cinders and Smoke: A Mile by Mile Guide to the Durango & Silverton Narrow Gauge Railroad.* 7th ed. Lakewood, Colo.: Western Guideways, 1995.
102. Petersen, David. *Among the Aspen: Life in an Aspen Grove.* Flagstaff, Ariz.: Northland Press, 1991.
103. Powell, Colin L., with Joseph E. Persico. *My American Journey.* New York: Random House, 1995.
104. Rennicke, Jeff. *The Rivers of Colorado.* Billings, Mont.: Falcon Press, 1985.
105. Robbins, Roy M. *Our Landed Heritage: The Public Domain, 1776–1936.* Lincoln: University of Nebraska Press, 1962.
106. Roberts, Willow. *Stokes Carson: Twentieth-Century Trading on the Navajo Reservation.* Albuquerque: University of New Mexico Press, 1987.
107. Rollins, Philip A. *The Cowboys: An Unconventional History of Civilization on the Old-Time Cattle Range.* 1922. Reprint, Albuquerque: University of New Mexico Press, 1976.
108. Sandoz, Mari. *The Cattlemen from the Rio Grande across the Far Marias.* New York: Hastings House, 1958.
109. Sibert, Edgar H., and Ted S. McKee. *Railfan's Guide to Colorado.* Boulder: Pruett Publishing, 1982.
110. Smith, Duane A. *Guide to Historic Durango & Silverton.* Evergreen, Colo.: Cordillera Press, 1988.
111. ———. *Horace Tabor: His Life and Legend.* Boulder: Colorado Associated University Press, 1973.
112. ———. *Mesa Verde National Park: Shadows of the Centuries.* Lawrence: University Press of Kansas, 1988.
113. ———. *Rocky Mountain Boom Town: A History of Durango.* Albuquerque: University of New Mexico Press, 1980.
114. ———. *Sacred Trust: The Birth and Development of Fort Lewis College.* Niwot, Colo.: University Press of Colorado, 1991.
115. ———. *Song of the Hammer and Drill: The Colorado San Juans, 1860–1914.* Golden, Colo.: Colorado School of Mines Press, 1982.

116. Smith, Toby. *Kid Blackie: Jack Dempsey's Colorado Days.* Ouray, Colo.: Wayfinder Press, 1987.

117. Stegner, Wallace. *Beyond the Hundredth Meridian: John Wesley Powell and the Second Opening of the West.* Boston: Houghton Mifflin, 1954.

118. Sturtevant, William C., gen. ed. *Handbook of North American Indians. Vol. 10, Southwest,* edited by Alfonso Ortiz. Washington D.C.: Smithsonian Institution, 1983.

119. ———, gen. ed. *Handbook of North American Indians. Vol. 11, Great Basin,* edited by Warren L. d'Azevedo. Washington, D.C.: Smithsonian Institution, 1986.

120. Tiller, Veronica E. Velarde. *The Jicarilla Apache Tribe: A History, 1846–1970.* Rev. ed. Lincoln: University of Nebraska Press, 1993.

121. Trachtenberg, Alan. *The Incorporation of America: Culture and Society in the Gilded Age.* New York, Hill and Wang, 1982.

122. Twitchell, Ralph E. *The Leading Facts of New Mexico History.* Vol. 2. Albuquerque: Horn and Wallace, 1963.

123. Utley, Robert M., and Wilcomb E. Washburn. *The American Heritage History of the Indian Wars.* 1977. Reprint, New York: Bonanza Books, 1982.

124. Vorpahl, Ben M. *Frederic Remington and the West: With the Eye of the Mind.* Austin: University of Texas Press, 1972.

125. Weatherford, Gary D., and F. Lee Brown, eds. *New Courses for the Colorado River: Major Issues for the Next Century.* Albuquerque: University of New Mexico Press, 1986.

126. Wellman, Paul I. *A Dynasty of Western Outlaws.* New York: Bonanza Books, 1961.

127. Wentworth, Edward N. *America's Sheep Trails: History, Personalities.* Ames: Iowa State College Press, 1948.

128. West, Elliott. *The Saloon on the Rocky Mountain Frontier.* Lincoln: University of Nebraska Press, 1979.

129. Westphall, Victor. *Mercedes Reales: Hispanic Land Grants of the Upper Rio Grande Region.* Albuquerque: University of New Mexico Press, 1983.

130. White, Richard. *"It's Your Misfortune and None of My Own": A History of the American West.* Norman: University of Oklahoma Press, 1991.

131. Wister, Owen. *The Virginian: A Horseman of the Plains.* New York: Macmillan, 1902.

132. Wolle, *Muriel S. Stampede to Timberline: The Ghost Towns and Mining Camps of Colorado.* Chicago: Sage Books, 1949.

133. Yagoda, Ben. *Will Rogers: A Biography.* New York: Harper-Collins, 1993.

Pamphlets

134. Arizona Public Service Co. *Four Corners Coal-Powered Generating Station.* N.p.: Arizona Public Service Co., n.d.

135. Armstrong, David M. *Lions, Ferrets, and Bears: A Guide to the Mammals of Colorado.* Denver: Colorado Division of Wildlife, 1993.

136. Colorado Division of Wildlife. *Durango Fish Hatchery, Visitors' Center & Wildlife Museum*. N.p., n.d.
137. ———. *Living with Wildlife in Bear Country*. N.p, n.d.
138. ———. *Living with Wildlife in Coyote Country*. N.p., n.d.
139. ———. *Living with Wildlife in Lion Country*. N.p., n.d.
140. ———. *State Trust Lands, '96–'97*. Denver, n.d.
141. Crow Canyon Archaeological Center. *Crow Canyon Archaeological Center: 1996 Catalog*. Cortez, Colo.: Crow Canyon Archaeological Center, 1995.
142. Farmington Convention and Visitors Bureau. *Farmington, New Mexico, 1995 Area Guide*. Farmington: Farmington Convention and Visitors Bureau, n.d.
143. Fort Lewis College. *Fort Lewis College*. Durango: Fort Lewis College, Office of Admissions and Development, 1996.
144. Reeves, Tim, and Alan Nelson. *Birds of Morgan Lake: A Guide to Common Species*. Waterflow, N.M.: Arizona Public Service Co., n.d.
145. Rocky Mountain Bighorn Sheep Society. *Dilemma of the Rocky Mountain Bighorn Sheep*. Denver: Rocky Mountain Bighorn Sheep Society, n.d.
146. San Juan Coal Company. *Welcome to the La Plata Mine*. La Plata, N.M.: San Juan Coal Co., n.d.
147. San Juan County Museum Association. *Salmon Ruin*. Bloomfield, N.M.: San Juan County Museum Association, n.d.
148. U.S. Bureau of Land Management. *Bisti Wilderness*. N.p., n.d.
149. ———. *Colorado Recreation Guide*. N.p.: U.S. Department of the Interior, n.d.
150. ———. *De-Na-Zin*. N.p.: U.S. Department of the Interior, n.d.
151. ———. *Explore Anasazi Heritage Center*. N.p.: U.S. Department of the Interior, n.d.
152. ———. *Guide to Angel Peak*. N.p.: U.S. Department of the Interior, n.d.
153. ———. *Pueblitos of Dinetah*. N.p.: U.S. Department of the Interior, n.d.
154. U.S. Bureau of Reclamation. *Navajo*. N.p.: U.S. Department of the Interior, n.d.
155. U.S. Forest Service. *Carson National Forest*. Washington, D.C.: U.S. Department of Agriculture, 1991.
156. ———. *Chimney Rock Archaeological Area*. N.p, n.d.
157. ———. *Falls Creek Archaeological Area*. N.p., n.d.
158. ———. *Lizard Head Wilderness Area*. N.p., n.d.
159. ———. *Weminuche Wilderness*. N.p., n.d.
160. ———. *Your National Forest: San Juan and Rio Grande National Forests*. N.p., n.d.
161. U.S. Park Service. *Aztec Ruins Official Map and Guide*. Washington D.C.: U.S. Department of the Interior, n.d.
162. ———. *Hovenweep*. N.p., n.d.
163. ———. *Mesa Verde National Park Colorado*. N.p., n.d.
164. Navajo Nation. *Four Corners Monument Navajo Tribal Park*. Window Rock, Ariz.: Navajo Nation, n.d.

165. Ute Mountain Ute Tribe. *Ute Mountain Tribal Park*. Towaoc, Colo.: Ute Mountain Ute Tribe, n.d.

Articles

Many of the articles in this section are from Pioneers of the San Juan Country, compiled by the Daughters of the American Revolution, Sarah Platt Decker Chapter, Durango (Colorado Springs: Out West Printing and Publishing, 1942–61; reprint, Bountiful, Utah: D.A.R. and Family History Publishers, 1995). Cited below as PSJC, the work originally appeared in four volumes but is now available in a single volume.

166. Ayres, Mary C. "Stockmen—Heathers and Ent." *PSJC*.
167. "Bayfield and the Pine River Valley." *PSJC*.
168. Bennett, T. Ralph. "Country Store Keepers: The Curtet Brothers." *PSJC*.
169. Bryce, Mrs. John. "A Lost Cemetery." *PSJC*.
170. ———. "Florida Mesa History." *PSJC*.
171. Cheasebro, Margaret. "For an Eerie Halloween, Spend the Night at Bisti." *Durango Herald Cross Currents Magazine*, October 20, 1995.
172. Cooper, Raymond H. "A History of San Juan County." *PSJC*.
173. Cornelius, Oliver F. "How Lady Luck Guided the Pick of Some Miners." *PSJC*.
174. ———. "Opening of the Ute Strip." *PSJC*.
175. ———. "Auxiliary Railroads in the San Juan Basin." *PSJC*.
176. Crum, Josie Moore. "The Rio Grande Southern Railroad." *PSJC*.
177. Daniels, Helen Sloan. "Lo, the Poor Indian." *PSJC*.
178. Eddy, Mark. "Ghost Grizzlies." *Denver Post Empire Magazine*. November 5, 1995.
179. Elliott, Nancy. "A Rifle for Home." *PSJC*.
180. Etheridge, Georgeanna. "Allison: The Neighborly Town." *PSJC*.
181. Heffernan, Frances Keegan. "Two Pioneer Families." *PSJC*.
182. Hemingway, Bill. "The Story of Cortez." *PSJC*.
183. Holmes, Bob. "The Big Importance of Little Towns on the Prairie." *National Wildlife*, June/July 1996.
184. Kunde, Karen. "Kroegers Shape Durango." *Newcomers Guide III: A Supplement to the Durango Herald*. June 16, 1996.
185. Lynch, June. "Pagosa! Pagosa! Healing Waters! A Legend from Grateful Utes." *PSJC*.
186. "Lynching of Bert Wilkinson." *PSJC*.
187. Marker, Nell. "Ignacio, the Town That Was Named for a Ute Chief." *PSJC*.
188. Netherton, Florence W. "Durango's First Newspaper." *PSJC*.
189. O'Keefe, Joanne. "Dual-Purpose Plant Will Make Steam and Electricity." *Durango Herald Cross Currents Magazine*, January 19, 1996.
190. Rasch, Philip J. "Feuding in Farmington." *New Mexico Historical Review*, July 1965.
191. Rhodes, Edith. "Bondad." *PSJC*.
192. Ricketts, Orval. "Frontier Farmington." *New Mexico*, September 1942.

193. Sadler, Lou P., and Estelle M. Camp. "Some Eminent People Who Have Lived in the San Juan Country." *PSJC*.
194. "San Juan Pioneers." *PSJC*.
195. Searcy, Helen M., and Estelle M. Camp. "A. P. Camp and the First National Bank." *PSJC*.
196. ———. "Otto Mears." *PSJC*.
197. ———. "The Military." *PSJC*.
198. Simms, J. Denton. "Emmet Wirt: Pioneer Extra-Ordinary." *PSJC*.
199. Smith, Duane A. "Willa Cather Loved Southwest Colorado." *Durango Herald Cross Currents Magazine*, May 3, 1996.
200. Uroda, Deborah. "Plan Could Return Healthy Forests and Profits." Durango Herald, October 20, 1995.
201. Weller, Robert. "Animas–La Plata Project Nearer to Reality." *Denver Post*, October 1, 1995.
202. West, George E. "The Oldest Range Man." *PSJC*.

Dictionaries, Encyclopedias, and Other Reference Works

203. American Association for State and Local History. *National Register of Historic Places, 1966–1991*. Nashville: American Association for State and Local History, 1992.
204. Beck, Warren A., and Ynez D. Haase. *Historical Atlas of the American West*. Norman: University of Oklahoma Press, 1989.
205. Benchmark Maps. *New Mexico Road and Recreation Atlas*. Berkeley: Benchmark Maps, 1995.
206. Bright, William. *Colorado Place Names*. Boulder: Johnson Publishing, 1993.
207. Carter, Jack L. *Trees and Shrubs of Colorado*. Boulder: Johnson Publishing, 1988.
208. Chamblin. Thomas S., ed. *The Historical Encyclopedia of Colorado*. Vol. 2. Rev. ed. Denver: Colorado Historical Association, 1975.
209. Colorado Scenic and Historic Byways Commission. *Discover Colorado: Colorado's Scenic and Historic Byways*. Denver: Colorado Scenic and Historic Byways Commission, n.d.
210. Colorado State Board of Stock Inspectors. *Colorado Brand Book, 1982*. Denver: Colorado Department of Agriculture, 1982.
211. Dawson, J. Frank. *Place Names in Colorado, Why 700 Communities Were So Named*. Denver: J. F. Dawson Publishing, 1954.
212. DeLorme Mapping. *Colorado Atlas and Gazetteer*. Freeport Maine: DeLorme Mapping, 1995.
213. ———. *Map'n'Go: North American Atlas and Gazetteer*. Freeport, Maine: DeLorme Mapping, 1994.
214. Eichler, George R. *Colorado Place Names: Communities, Counties, Peaks, Passes*. Boulder: Johnson Publishing, 1977.
215. *Farmington Fact Book*. City of Farmington, 1989.

216. Frazer, Robert W. *Forts of the West*. Norman: University of Oklahoma Press, 1972.
217. Funk and Wagnalls. *Microsoft Encarta*. Microsoft Corp., 1994.
218. Halliwell, Leslie, and John Walker, eds. *Halliwell's Film Guide*. New York: Harper & Row, 1995.
219. Helmuth, Ed, and Gloria Helmuth. *Passes of Colorado: An Encyclopedia of Watershed Divides*. Boulder: Pruett Publishing, 1994.
220. Klein, Barry T. *Reference Encyclopedia of the American Indian*. 5th ed. West Nyack, N.Y.: Todd Publications, 1990.
221. Lamar, Howard R., ed. *Reader's Encyclopedia of the American West*. New York: Harper & Row, 1977.
222. Low, W. Augustus, and Virgil A. Clift, eds. *Encyclopedia of Black America*. New York: McGraw-Hill, 1981.
223. McLoughlin, Denis. *Wild and Woolly: An Encyclopedia of the Old West*. Garden City, N.Y.: Doubleday, 1975.
224. Moritz, Charles, ed. *Current Biography Yearbook, 1988*. New York: H. W. Wilson Co., 1989.
225. Nelson, Ruth A. *Handbook of Rocky Mountain Plants*. Tucson: Dale Stuart King, 1969.
226. New Mexico Livestock Board. *New Mexico Brand Book*. Albuquerque: New Mexico Livestock Board, 1994.
227. *New Mexico State and Federal Natural Resources Recreation Map*. N.p., n.d.
228. Noel, Thomas J., Paul F. Mahoney, and Richard E. Stevens. *Historical Atlas of Colorado*. Norman: University of Oklahoma Press, 1993.
229. Odd, Gilbert E. *Encyclopedia of Boxing*. New York: Crescent Books, 1983.
230. Office of the Federal Register. *United States Government Manual: 1995/96*. Washington D.C.: National Archives and Records Service, 1995.
231. O'Neal, Bill. *Encyclopedia of Western Gunfighters*. Norman: University of Oklahoma Press, 1979.
232. Pearce, T. M. *New Mexico Place Names*. Albuquerque: University of New Mexico Press, 1965.
233. Peattie, Donald C. *A Natural History of Western Trees*. Boston: Houghton Mifflin, 1991.
234. Ploski, Harry A., and James Williams, eds. *The Negro Almanac: A Reference Work on the Afro-American*. 4th ed. New York: John Wiley and Sons, 1983.
235. Preston, Richard J. *Rocky Mountain Trees*. 2d ed. Ames, Iowa: Iowa State College Press, 1947.
236. Rocky Mountain Association of Geologists. *Geologic Atlas of the Rocky Mountain Region, United States of America*. Denver: Rocky Mountain Association of Geologists, 1972.
237. Shearer, Benjamin F., and Barbara S. Shearer. *State Names, Seals, Flags and Symbols: A Historical Guide*. New York: Greenwood Press, 1994.
238. Shearer Publishing. *The Roads of New Mexico*. Fredericksburg, Tex., 1990.
239. Smith, Colin. *Collins Spanish-English English-Spanish Dictionary*. 2nd ed. Glasgow, G.B.: Collins Publishers, 1988.

240. Stanley, F. [Francis Louis Crocchiola]. *Desperadoes of New Mexico.* Denver: World Press, 1953.

241. Straub, Deborah A., ed. *Contemporary Authors.* Detroit: Gale Research Co., 1987.

242. Tuska, Jon, and Vicki Piekarski. eds. *Encyclopedia of Frontier and Western Fiction.* New York: McGraw-Hill, 1983.

243. Ungnade, Herbert E. *Guide to New Mexico Mountains.* 6th ed. Albuquerque: University of New Mexico Press, 1983.

244. U.S. Geological Survey. *Geologic Atlas of the United States.* Washington, D.C.: U.S. Department of the Interior, 1899–1916.

245. Waldman, Carl. *Atlas of the North American Indian.* New York: Facts on File, 1985.

246. ———. *Who Was Who in Native American History.* New York: Facts on File, 1990.

247. Williams, Jerry L., and Paul E. McAllister, eds. *New Mexico in Maps.* Albuquerque: University of New Mexico Technology Application Center, 1986.

Miscellaneous Sources

248. Benham, Jack L. *100 Years in the San Juans.* Ouray, Colo.: Bear Creek Publishing, 1981.

249. Box, Eddie, Jr., and Betty Box. "Southern Ute History." *Homepage of Eddie and Betty Box.* <http://animas.frontier.net/~ebox/>. October 5, 1996.

250. Colorado Division of Wildlife. *State Wildlife Area Information for Southwest Colorado.* Durango: Colorado Division of Wildlife, n.d.

251. Crofutt, George A. *Crofutt's Grip-Sack Guide of Colorado,* 1885. Omaha: Overland Publishing, 1885.

252. Denham A. H., and J. S. Banks. *History and Program at San Juan Basin Research Center.* Fort Collins: Colorado State University, 1988.

253. Gutierrez, Donna. "Description of the Existing Environment Topography." San Juan Basin Regional Uranium Study, Working Paper No. 17. Albuquerque: U.S. Department of the Interior, n.d.

254. La Plata County Clerk. *Minutes of the Meeting of Board of County Commissioners, October 4, 1881.* Durango, Colorado.

255. San Juan Basin Area Vocational-Technical School. *San Juan Basin Technical School, 1995–98 Catalog.* Cortez: San Juan Basin Area Vocational-Technical School, 1994.

256. Southwest Colorado Film Commission. *Colorado: It Helps to Have Friends in High Places.* Durango: Southwest Colorado Film Commission, 1992.

257. U.S. Bureau of Reclamation. *Colorado River Storage Project.* N.p., 1983.

258. ———. *Dolores Project.* N.p., 1983.

259. ———. *Florida Project.* N.p., 1982.

260. ———. *Hammond Project.* N.p., 1980.
261. ———. *Mancos Project.* N.p., 1983.
262. ———. *Navajo Indian Irrigation Project.* N.p., 1973.
263. ———. *New Mexico Bureau of Reclamation Projects.* N.p., 1983.
264. ———. *Pine River Project.* N.p., 1983.
265. ———. *What Is Reclamation?* N.p., n.d.
266. U.S. Department of the Interior. *National Performance Review Laboratory: Navajo Reservoir Management Area.* Washington, D.C.: U.S. Department of the Interior, n.d.
267. U.S. Forest Service. *Celebrate National Forests, 1891–1991.* Washington, D.C.: U.S. Department of Agriculture, 1990.